Michael Schöbel

Residual Stresses and Thermal Fatigue in Metal Matrix Composites

Michael Schöbel

Residual Stresses and Thermal Fatigue in Metal Matrix Composites

A comparative study of stress measurements by neutron and synchrotron diffraction combined with tomography

Südwestdeutscher Verlag für Hochschulschriften

Impressum/Imprint (nur für Deutschland/only for Germany)
Bibliografische Information der Deutschen Nationalbibliothek: Die Deutsche Nationalbibliothek verzeichnet diese Publikation in der Deutschen Nationalbibliografie; detaillierte bibliografische Daten sind im Internet über http://dnb.d-nb.de abrufbar.
Alle in diesem Buch genannten Marken und Produktnamen unterliegen warenzeichen-, marken- oder patentrechtlichem Schutz bzw. sind Warenzeichen oder eingetragene Warenzeichen der jeweiligen Inhaber. Die Wiedergabe von Marken, Produktnamen, Gebrauchsnamen, Handelsnamen, Warenbezeichnungen u.s.w. in diesem Werk berechtigt auch ohne besondere Kennzeichnung nicht zu der Annahme, dass solche Namen im Sinne der Warenzeichen- und Markenschutzgesetzgebung als frei zu betrachten wären und daher von jedermann benutzt werden dürften.

Verlag: Südwestdeutscher Verlag für Hochschulschriften GmbH & Co. KG
Heinrich-Böcking-Str. 6-8, 66121 Saarbrücken, Deutschland
Telefon +49 681 37 20 271-1, Telefax +49 681 37 20 271-0
Email: info@svh-verlag.de

Approved by: Wien, TU, Diss., 2011

Herstellung in Deutschland:
Schaltungsdienst Lange o.H.G., Berlin
Books on Demand GmbH, Norderstedt
Reha GmbH, Saarbrücken
Amazon Distribution GmbH, Leipzig
ISBN: 978-3-8381-2899-3

Imprint (only for USA, GB)
Bibliographic information published by the Deutsche Nationalbibliothek: The Deutsche Nationalbibliothek lists this publication in the Deutsche Nationalbibliografie; detailed bibliographic data are available in the Internet at http://dnb.d-nb.de.
Any brand names and product names mentioned in this book are subject to trademark, brand or patent protection and are trademarks or registered trademarks of their respective holders. The use of brand names, product names, common names, trade names, product descriptions etc. even without a particular marking in this works is in no way to be construed to mean that such names may be regarded as unrestricted in respect of trademark and brand protection legislation and could thus be used by anyone.

Publisher: Südwestdeutscher Verlag für Hochschulschriften GmbH & Co. KG
Heinrich-Böcking-Str. 6-8, 66121 Saarbrücken, Germany
Phone +49 681 37 20 271-1, Fax +49 681 37 20 271-0
Email: info@svh-verlag.de

Printed in the U.S.A.
Printed in the U.K. by (see last page)
ISBN: 978-3-8381-2899-3

Copyright © 2011 by the author and Südwestdeutscher Verlag für Hochschulschriften GmbH & Co. KG and licensors
All rights reserved. Saarbrücken 2011

This work is dedicated to
Heidemarie Knoblich
1947 – 2011

Abstract

This work is dealing with non-destructive testing of heterogeneous materials for fundamental research in materials science. Recently developed metal matrix composites emerging in the field of high performance heat sink materials are investigated concerning their internal stresses and thermal fatigue damage mechanisms under simulated operation conditions. New methods like synchrotron and neutron diffraction are combined with synchrotron tomography to enable new insights of composites. Particle as well as monofilament reinforced metals are investigated covering two main composite architectures. Diffraction gives information on the micro stresses in the composite during thermal cycling due to the mismatch in thermal expansion and in the Young's moduli between matrix and reinforcement. Matrix deformation and damage as well as delamination at the interfaces are caused by those stresses and correlated relaxation can be identified by diffraction owing to a reduction of the stress amplitude. Tomographic imaging of the internal structure is achieved in situ during thermal cycling revealing thermal fatigue damage. Pore evolution and crack propagation are visualized in 3D in the bulk of the material.

Initial voids are identified in particle reinforced aluminum with high particle volume fractions of SiC or diamond (PRM: Al and AlSi7(Mg)/SiC/60-70p or Al and AlSi7/CD/60p) and big CTE mismatch (ΔCTE = 20 and 25 ppm/K, respectively), which are formed by a shrinking matrix metal in a densely packed particle preform. These voids change their volume fraction during thermal cycling (RT – 350°C) by stress induced visco-plastic matrix deformation. An anomalous CTE decrease is caused by internal accommodation of the expanding matrix into those voids, which shrink during heating and reopen during cooling again. Eutectic Si segregation in the matrix alloy changes the reinforcement architecture from isolated particles embedded in a matrix into an interpenetrating reinforcement network by bridging the particles thus embedding the metal matrix. The long term stability of heat sinks for power electronic components using a 3D reinforcement architecture forming an interpenetrating composite is significantly improved concerning delamination during heating.

SiC monofilament or W fiber reinforced Cu (MFRM: Cu and CuCr1Zr/SiC or W/10 - 50m) combine the low CTE of the fiber with the high thermal conductivity of the Cu matrix. Sufficient interface bonding quality is important for the long term stability of the MFRM and is improved by Ti coated SiC-monofilaments. Weak bonding causes delamination at the fiber-matrix interfaces, and too strong bonding causes matrix damage by interfacial shear stresses during thermal cycling (RT - 550°C). Fiber cracks could be identified in the brittle SiC monofilaments. These cracks originated from production and cause accumulating elongation by thermal fatigue damage. Those cracks could not be observed in the more ductile W fibers. W monofilament reinforced Cu has a strong bonding even without any interface treatment. Cu/W/20m exhibit good bonding at high micro stress amplitudes, but severe matrix plastification occurs owing to the limited strength of pure Cu. In monofilament reinforced copper composites the interface bonding strength has to be balanced between stress induced interface delamination (if stress > bonding strength) and matrix shear strength (if stress > matrix shear strength) to achieve a good long term stability under intended operation conditions of divertor components in fusion reactors.

Content

1. Heat sink materials	1
1.1. Introduction	1
1.2. Material requirements	2
1.3. Applications	3
1.3.1. Power electronics	3
1.3.2. Fusion reactor systems	6
References	11
2. Metal matrix composites	12
2.1. MMC categories	12
2.2. Material properties	14
2.2.1. Thermal conductivity	14
2.2.2. Thermal expansion	16
2.3. Internal stresses	17
2.3.1. Mechanical load	17
2.3.2. Thermal load	20
2.3.3. Eshelby model	21
References	24
3. Investigated MMC	25
3.1. Particle reinforced metals (PRM)	25
3.1.1. Al-SiC	25
3.1.2. Diamond reinforced metals	27
3.1.3. PRM production process	29
Gas pressure infiltration	29
Squeeze casting	30
3.1.4. Particles	30
3.2. Monofilament reinforced metals (MFRM)	31
3.2.1. SiC-fiber reinforced copper	32
3.2.2. W-wire reinforced copper	33
3.3. MFRM production process	34
3.4. Monofilaments	36
3.5. Investigated samples	38
References	41
4. Aims of investigations	43

5. Experimental background	44
5.1. Diffraction	44
5.2. Radiation types	49
5.2.1. X-rays	50
5.2.2. X-ray sources	52
5.2.3. X-ray detectors	58
5.2.4. Neutrons	63
5.2.5. Neutron sources	65
5.2.6. Neutron detectors	67
5.3. Radiation properties	68
Comparison	70
5.4. Monochromatic diffraction on a polycrystalline material	71
5.5. Stress analysis by diffraction methods	73
5.6. Diffraction experiment	79
5.7. Residual stresses in the diffraction pattern	84
5.8. Systematic errors	85
5.9. Tomography / Imaging	89
5.9.1. Fundamentals	89
5.9.2. Tomography experiment	90
5.9.3. Post processing	92
References	93
6. Discussion of thermal fatigue damage in MMC	94
6.1. High volume fraction particle reinforced composites (PRM)	94
6.2. Monofilament reinforced composites (MFRM)	96
6.3. Thermal fatigue damage types	97
References	100
7. Obtained results	101
Conclusion for MMC design	102

Reinforcement architecture and thermal fatigue in
diamond particle reinforced aluminum,
Acta Materialia, 2010.

    103

Internal stresses and voids in SiC particle reinforced
aluminum composites for heat sink applications,
Composite Science and Technology, 2011.

    113

Thermal fatigue in monofilament reinforced copper
for heat sink applications in divertor elements,
Journal of Nuclear Materials, 2011.

    123

1. Heat sink materials

1.1. Introduction

This work is based on a part of the contribution of TU Vienna to the EU Integrated Project "ExtreMat" (Materials for Extreme Environments) dealing with non-destructive testing of novel heat sink materials. The ExtreMat project [1] was created by Max-Planck-Institut für Plasma Physik (IPP), Garching, Germany. Professor Dr. Harald Bolt, head of the institute of materials science at IPP, initiated the science and industrial consortium. After three years of preparation, the project started on 1st of December 2004. The goal of ExtreMat was to develop and to applicate new materials for extreme environments. Several materials have been developed and tested for applications such as power electronic modules for hybrid vehicles, high precision laser optics for space applications and new fusion reactor systems. At the same time ExtreMat implemented an integrated project in the 6th European framework program with a budget of 35 M€ and 38 European partners involved.

This thesis describes the cooperation of TU Vienna with EMPA Thun [2], IPP Garching [3], EPFL Lausanne [4], DLR Köln [5], IFAM Dresden [6], Plansee Reutte [7]. Its aim was to identify and to evaluate residual stresses and thermal fatigue damage in metal matrix composites for heat sink applications. Several non-destructive diffraction techniques have been applied for in/ex situ stress measurements supported by computed tomography.

The investigation of diamond reinforced metals was held comparable to previous work on silicon carbide reinforced aluminum composites (AlSiC). These diamond reinforced composites are developed to be replaced AlSiC as high conducting heat sink material with an improved performance in application. New possibilities that were achieved by highly brilliant synchrotron X-ray sources [8] and high flux neutron sources [9, 10] allowed new insights into micro stresses, their origin and propagation under application conditions. In situ synchrotron computed tomography (SCT) enables to visualize thermal fatigue damage, voids and plastic deformation in the matrix metal during thermal cycling. Thus, questions left open in previous work on AlSiC [11] could be answered. The conclusions of AlSiC on new diamond reinforced composites [12] were proved in their applicability.

Monofilament reinforced copper composites had to be investigated under operation conditions [13]. These fiber reinforced composites are developed for heat sink applications in divertor elements of fusion reactor systems. In situ neutron diffraction (ND) was performed for stress measurements supported by synchrotron diffraction (SD) and SCT. Tungsten and SiC fibers were compared. Pure copper as well as copper-chromium-zircon alloy (CuCr1Zr) were tested. Diffraction experiments delivered information on the internal stresses which represent the driving force of thermal fatigue. Complementary tomography could show such fatigue damage types in 3D.

New methods applied to new composite materials gave relevant information on internal architectures and their effects on residual stresses and thermal fatigue damage, particularly in composites with big coefficient of thermal expansion (CTE) mismatch. The combination of in situ diffraction and in situ tomography turned out to be a useful method to investigate these structures under simulated application conditions. Three more comprehensive publications (Tab. 1.1) are included in this work, each representative for the composite type investigated. Similar experimental techniques were applied for material characterization of particle reinforced metals (PRM) such as: Al-SiC and Al-carbon diamond (Al-CD), as well as monofilament reinforced metals (MFRM) such as: Cu-SiC and Cu-W.

Publication	Title	Magazine	Authors
Diamond reinforced Aluminium	Reinforcement architectures and thermal fatigue in diamond particle reinforced aluminium	Acta Materialia, Volume 58, Issue 19, November 2010, Pages 6421-6430	M. Schöbel[1], H.P. Degischer[1], S. Vaucher[2], M. Hofmann[3], P. Cloetens[4]
SiC reinforced Aluminium	Internal stresses and voids in SiC particle reinforced aluminum composites for heat sink applications	Composites Science and Technology, Volume 71, Issue 5, 22 March 2011, Pages 724-733	M. Schöbel[1], W. Altendorfer[1], H.P. Degischer[1], S. Vaucher[2], T. Buslaps[4], M. Di Michiel[3], M. Hofmann[3]
Monofilament reinforced Copper	Thermal fatigue in monofilament reinforced copper for heat sink applications in divertor elements	Journal of Nucleas Materials, Volume 409, Issue 3, February 2011, Pages 225-234	M. Schöbel[1], J. Jonke[1], H.P. Degischer[1], V. Paffenholz[5], A. Brendel[5], R.C. Wimpory[6], M. Di Michiel[4]

[1]Institute of Materials Science and Technology, Vienna University of Technology, Karlsplatz 13, A-1040 Vienna, Austria
[2]Advanced Materials Processing, EMPA – Swiss Federal Laboratories for Materials Science and Technology, Feuerwerkstrasse 39, CH-3602 Thun, Switzerland
[3]Forschungsneutronenquelle Heinz Maier-Leibnitz, Lichtenbergstr.1, D-85747 Garching, Germany
[4]European Synchrotron Radiation Facility, 6 Rue Jules Horowitz, F-38043 Grenoble, France
[5]Max-Planck-Institut für Plasmaphysik, Boltzmannstrasse 2, D-85748 Garching, Germany
[6]Helmholtz Zentrum Berlin, Hahn Meitner Platz 1, D-14109 Berlin, Wannsee, Germany

Tab. 1.1.: The publications included in this work.

1.2. Material requirements

In recent years the importance of new types of heat sink materials became evident by an increasing amount of high power density situations in high temperature and / or small scale applications [1]. High and mostly heterogeneous thermal load on components under operation conditions results into a reduction of long term stability as a consequence of thermal fatigue damage. Future demands will be high and need invention of new heat sink materials and further developments of those already known. For this purpose composites have to be developed which enable to combine properties of two entirely different materials [14]. In metal matrix composites the high TC of a metal matrix is combined with the low CTE of a stiff ceramic reinforcement. Often a heat sink with a suitable CTE is necessary so that damage and delamination of two components with CTE mismatch can be avoided. The application of a particle or fiber reinforced composite interlayer transports mismatch stresses from the macroscopic interface into the bulk volume of the composite. The stresses are aggregated by a bigger surface region between the matrix and the reinforcement. However, these particle / fiber-matrix micro stresses will cause internal thermal fatigue damage in the composite [15], if the bonding strength is too low and/or the temperature change and the CTE mismatch too high. Metal matrix composites (MMC) were investigated regarding their interface strength and reinforcement architecture which effect such internal stresses during thermal cycling.

1.3. Applications

High thermal conductivity (TC) of a heat sink material guarantees sufficient temperature flux from the heat source into the cooling medium [16]. A suitable (low) coefficient of thermal expansion (CTE) is required to increase the long term stability during cycling thermal load under service conditions. MMC combine the high TC of a metal with low CTE of a ceramic reinforcement [17]. In the following passage two major applications of the materials investigated are described.

1.3.1. Power electronics

Power electronic converters are used wherever a change in voltage, current or frequency is needed [18]. The conversion is performed by semiconductor switching devices. The power range extends from a few watt in television, computers or battery chargers up to tenths of megawatt for high power applications, such as production of power pulses for particle and plasma physics, high-voltage in solid-state Tesla coils and coil-guns or generators for induction motors in hybrid vehicles and railway traction. The Insulated Gate Bipolar Transistor module (IGBT) is a fairly recent invention, meanwhile available in third generation (Fig. 1.1).

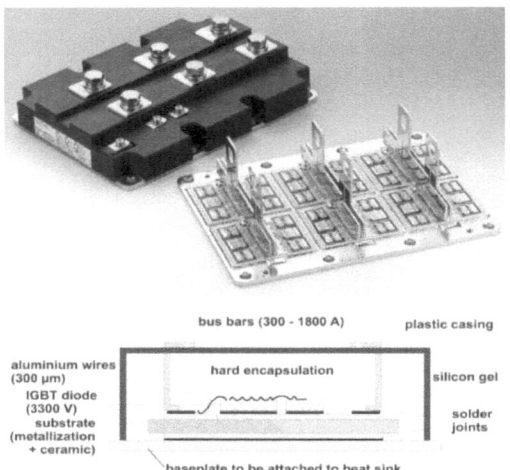

Fig. 1.1.: An IGBT (Insulated Gate Bipolar Transistor) module with sketch of its cross section below [19].

Discovered experimentally by B. Jayant Baliga in 1979 [20], the IGBTs have been developed to replace the MOSFETs and GTO-thyristors for power electronic applications. Fig. 1.2 shows a circuit diagram of an IGBT with NPT (Non Punsh Through) structure [18].

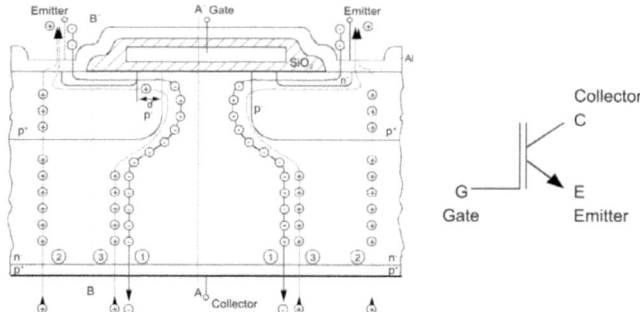

Fig. 1.2.: The current through an IGBT in non punsh through geometry [18]. Circuit diagram on the right.

The IGBT module consists of semiconductor layers which are epitaxically grown from a highly doped p^+-silicon substrate base layer. On top of the low doped n^--drift region two p-cathode tanks with highly doped n^+-islands, diffusion grown between gate and emitter. Operating the IGBT with its typical npnp-structure, a constant voltage U_{CE} (~ 6 kV) is applied between collector and emitter. If a gate voltage U_{GE} is applied, usually between 12 and 15 V, a conducting n-channel is produced in the cathode tank. Electrons can pass from the emitter through the n^+-p^- region into the epitactic. On the bottom, the highly doped p^+-substrate induces holes in the epitactic producing a conducting electron-hole plasma. The holes move not only through the drift region to the emitter but also, beside the n^+-region, directly from the collector to the emitter. The n^--drift region is flooded by charge carriers, and the collector-emitter voltage is reduced. The recombination of the conducting plasma in the epitaxic n^--layer produces high switching losses in IGBTs compared to MOSFETs. Therefore, the IGBT modules are preferably used for high voltages > some 100 V and low frequencies from 1 to 20 kHz. In their typical operation region new IGBTs achieves ~ 7 kV and 3 kA competing with high power MOSFETs and GTO-thyristors.

The high power densities (~ 5×10^5 W/cm^2) in the IGBT modules generate heat which has to be dissipated from the Si chips into a heat sink [17]. An advanced thermal management is required which has to fulfill high cooling performance under long term operation conditions. The heat produced in an IGBT has to pass through several layers of different materials into an air cooled heat sink (Fig. 1.3).

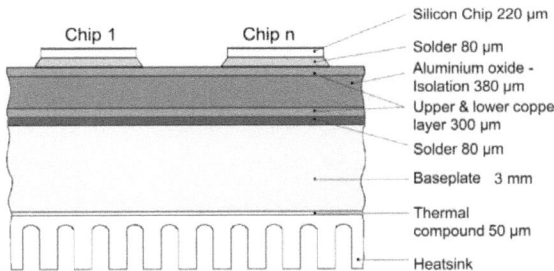

Fig. 1.3.: The layers of the electronic packaging in an IGBT [18].

This simplified view illustrates how the several components contribute to the thermal gradient from the chip to the heat sink in an IGBT module. The chips are soldered on an Al_2O_3 electrical insolating interlayer (~ 380 µm), soldered on a baseplate (~ 3 - 5 mm) which is thermally contacted to an air cooled heat sink. The baseplate acts as high conducting heat spreader which distributes the temperature from the small chips over a big area into the heat sink, but also avoid thermal stresses within the ceramic chips. To achieve uniformly good thermal contact on the interface, the baseplate is formed convex and pressed against the plane surface of the heat sink (Fig. 1.4). The elastic preload improves homogeneous thermal contact at the interface to increase the stability of the component under operation conditions.

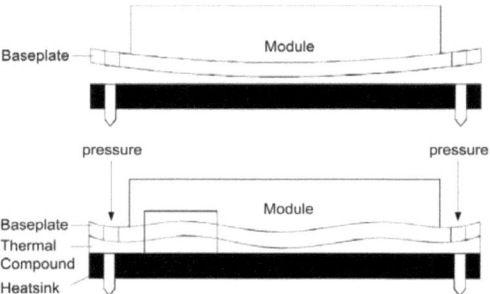

Fig. 1.4.: Baseplate with attached chips before (above) and after (below) being mounted on the heat sink [18].

The ceramic aluminum oxide insulator layer is soldered on the baseplate at ~ 350 °C. This solder region suffers high thermo-mechanical mismatch stresses from production as well as under operation conditions. Stress gradients and CTE mismatch generates delamination during a cycling thermal load of the chips between -50 °C up to 150 °C. Thermal fatigue damage at this chip-baseplate interface reduces the regional thermal conductivity below the ceramic chips. In order to avoid these mismatch-stresses a baseplate material with a low coefficient of thermal expansion (CTE) suitable to the Al_2O_3 ceramic layer has to be applied. The baseplates are usually made of CuMo or CuW alloys with a low CTE and good thermal conductivity. Long term performance can be improved when MMC are used as baseplate materials, consisting of an aluminum matrix alloy reinforced with high volume fractions of SiC particles [11, 16, 17]. Further performance increase of high power IGBT modules requires new baseplate materials with superior thermal properties. Therefore particle reinforced composites are developed as highly conducting heat spreader in such applications.

1.3.2. Fusion reactor systems

The ever-increasing demand on energy requires new methods of energy extraction. Over the past decades regenerative sources had to be optimized and new ones developed in order to supply a continuously growing human population with energy all over the world. Regarding that recently developed water, solar and wind power stations are still limited in their physical and economic efficiency, they will not be able to displace non-regenerative conventional fossil and nuclear (fission) energy sources for the amount of energy to be covered. New solutions have to be found as nuclear fusion is offering the most promising possibilities for the future [21].

The main achievement of fusion compared to fission is the good availability of the raw materials. Deuterium and lithium (needed for Tritium breeding) will be available on earth for millions of years. Moreover, uncontrolled run off scenarios of the reaction under operation conditions are impossible, and the lack of long term activation of residuals from the nuclear reaction excludes any storage problem of radioactive waste, well-known from fission reactors.

In fusion, a deuteron (D) and trition (T) is reacting to an alpha particle (He core) and a neutron (n) (Fig. 1.5).

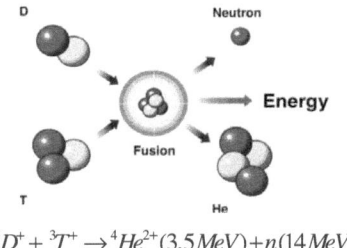

$$^2D^+ + {}^3T^+ \rightarrow {}^4He^{2+}(3,5 MeV) + n(14 MeV)$$

Fig. 1.5.: Fusion reaction with its process equation below [22].

This exotherm process generates ~ 17.5 MeV per reaction by an energy relieve from the potential between nuclear particles (strong interaction) [23]. In order to overcome the repulsive electromagnetic force of the charged nuclei, a high activation temperature (energy) is required. The hydrogen plasma reacts at a density of ~ 10^{20} m^{-3} at energies of ~ 18 keV (200 Mio. K). No solid state material would withstand these extreme temperatures. Therefore thermal isolation of the plasma from the components of the reactor systems is essential [24]. The Tokamak reactor type [3, 25] uses three superimposed magnetic fields to stabilize the plasma in the center of the reaction chamber and avoids direct thermal contact with the inner wall (Fig. 1.6).

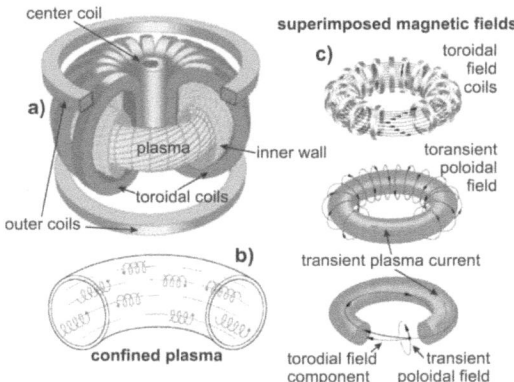

Fig. 1.6.: A schematic view of a Tokamak reactor (a) [3, 26]. The toridal field confines the plasma between the inner walls (b) to avoid direct thermal contact. The magnetic field is a superposition of several coil systems (c).

A complex arrangement of superconducting coils is responsible for the field generation in the plasma. The toroidal coils keep the plasma on its ring shaped trajectory around the center of the reactor. Large outer coils induce a tangential current in the plasma. The big vertical central coil acting as transformer produces a helical plasma flow around the surface of the enclosed plasma torus. This helical movement avoids turbulences between inner and outer regions in the plasma which would destabilize the plasma current under operation. In this geometry, the plasma serves as secondary winding of a transformer. A limitation of the current in the primary inner central coil demands a pulsed operation of the Tokamak reactor type.

The inner wall of the reaction chamber does not only suffer heavy thermal load but also high energy particle irradiation [24]. Fig. 1.7 shows the reaction chamber of JET (Joint European Torus) [27].

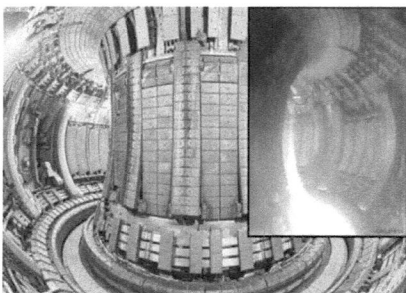

Fig. 1.7.: The reaction chamber of the JET torus with a view on the plasma reaction under operation [27].

The inner wall of the reaction chamber is cladded with tungsten as plasma facing material due to its high temperature resistance. In order to transport the heat from the chamber surface into a cooling medium, a complex thermal management using advanced materials has to be applied to ensure efficient temperature dissipation and long term stability under operation conditions [24]. Fiber reinforced composites are favored materials due to their reduced embrittlement under heavy irradiation by crack growth inhibition at the fiber-matrix interfaces [28]. A cross section through the toroidal reaction chamber shows the field inside the walls in which the plasma is captured (Fig. 1.8).

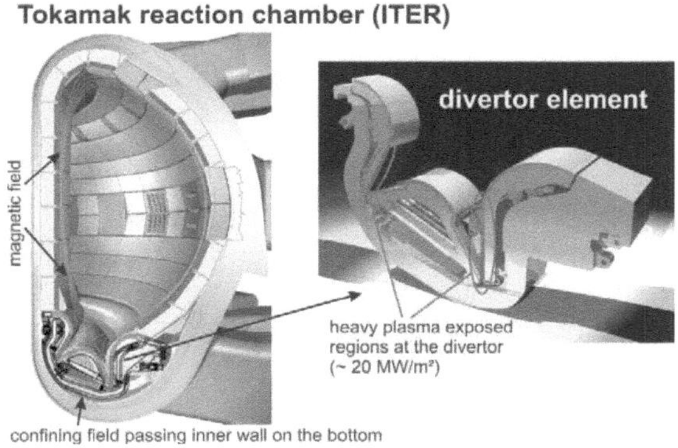

Tokamak reaction chamber (ITER)

magnetic field

divertor element

heavy plasma exposed regions at the divertor (~ 20 MW/m²)

confining field passing inner wall on the bottom

Fig. 1.8.: Section through the ITER's (International Thermonuclear Experimental Reactor) reaction chamber with field lines inside the inner walls [29, 30]. The divertor on the bottom collects the residuals of the nuclear reaction where the field penetrates the inner wall.

The highest plasma particle flux into the walls is located at the bottom of the chamber where the magnetic field penetrates the surface of the component [24]. The divertor elements placed in these highly loaded regions clean the plasma by collecting the residuals from the nuclear reaction. Thermal loads of up to 20 MW/m² require a high performance cooling solution. In the divertor a plasma facing tungsten plate has to be contacted thermally well to an actively cooled CuCrZr heat sink (Fig. 1.9).

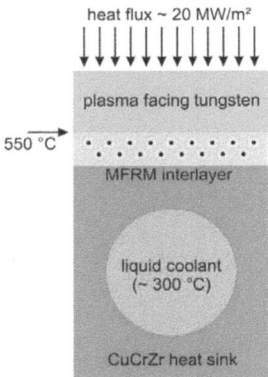

Fig. 1.9.: The plasma facing tungsten plate attached to an actively cooled CuCr1Zr heat sink [31]. A MFRM (Monofilament Reinforced Metal) interlayer is applied in a region with operating temperatures of ~ 550 °C and 20 MW/mK thermal load.

The required pulsed operation of the Tokamak reactor type causes high thermal mismatch stresses at changing temperatures between W and CuCrZr (ΔCTE ~ 12 ppm/K) [32]. The stresses cause delamination under service and, as a consequence, a degradation of the thermal contact important for cooling. In order to reduce these stresses, fiber reinforced copper composites are developed as an interlayer material with an intermediate CTE [31, 33]. SiC or W monofilaments decrease the CTE of the high thermally conducting Cu matrix in fiber direction and provide temperature strength.

References

[1] http://www.extremat.org
[2] http://www.empa.ch
[3] http://www.ipp.mpg.de
[4] http://www.epfl.ch
[5] http://www.dlr.de
[6] http://www.ifam.fraunhofer.de
[7] http://www.plansee.at
[8] http://www.esrf.fr
[9] http://www.frm2.tum.de
[10] http://www.helmholtz-berlin.de/
[11] Huber T, Degischer HP, Lefranc G, Schmitt T, Thermal expansion studies on aluminium-matrix composites with different reinforcement architecture of SiC particles, Comp. Sci. Tech. vol. 66, 2206-2217, 2006.
[12] Ruch PW, Beffort O, Kleiner S, Weber L, Uggowitzer PJ, Selective interface bonding in Al(Si)-diamond composites and its effect on thermal conductivity, Comp. Sci. Tech. vol. 66, 2677-2685, 2006.
[13] Peters P, Hemptenmacher J, Schurmann H, The fiber/matrix interface and its influence on mechanical and physical properties of Cu-MMC. in press, Comp. Sci. Tech. vol. 70, 1321-1329, 2010.
[14] Mortensen A, Kelly A, Zweben C, Comprehensive Composite Materials, vol. 3, Metal Matrix Composites, 541-547, Oxford 2000.
[15] Stinchcomb WW, Ashbaugh NE, Composites Materials: Fatigue and Fracture, vol. 4, ASTM, 1916 Race Street, Philadelphia, PA 19103, 1993.
[16] Zweben C, Metal-matrix composites for electronic packaging, J Metal, vol. 44, 15-23, 1992.
[17] Lefranc G, Degischer HP, Sommer HK, Mitic G. Al-SiC improves reliability of IGBT power modules, In: Massard T, ICCM12 proceedings, Paris, electronic support, 1999.
[18] http://www.semikron.com
[19] Huber T, Thermal expansion of Aluminium Alloys and Composites, PhD-thesis, Institute of Materials Science and Technology, TU-Vienna, 2003.
[20] Baliga BJ, Enhancement and Depletion Mode Vertical Channel MOS Gated Thyristors, Electronics Letters, vol. 15, 645-647, 1979.
[21] Bogush E, Bolt H, Chevalier A, Forty C, Gnesotto F, Heller R et al, Benefits of European industry from involvement in fusion, Fus. Eng. Des. vol. 63-64, 679-687, 2002.
[22] http://helian.net/blog
[23] Braams CM, Stott PE, Nuclear Fusion, Half a Century of Magnetic Confinement Fusion Research, Institute of Physics Publishing, Bristol and Philadelphia, 2002.
[24] Becvar E, Aspekte der Kernfusionsforschung, Verlag der Österreichischen Akademie der Wissenschaften, Informationstagung April 1986.
[25] http://fti.neep.wisc.edu
[26] http://www.wikipedia.org/wiki/tokamak
[27] http://www-fusion-magnetique.cea.fr
[28] Brendel A, Popescu C, Köck T, Bolt H, Promising composite heat sink material for divertor of future fusion reactors, Jour. Nuc. Mat. 367-370, 2007.
[29] Linsmeier C, Synchrotron- und neutronenbasierte Charakterisierung von Fusionsmaterialien, DGM Fachausschuss: Werkstoffuntersuchungen mit Strahlenlinien, Hamburg, April 2010.
[30] http://www.naka.jaea.go.jp
[31] Paffenholz V, Synthese und Charakterisierung von SiC$_f$/Cu-Matrix-Verbundwerkstoffe und ihre Anwendung in einem Modell einer Divertor-Komponente, PhD-thesis, IPP, TU-München, 2010.
[32] Linke J, Escourbiac F, Mazul IV, Nygren R, Rödig M, Schlosser J, Suzuki S, High heat flux testing of plasma facing materials and components - Status and perspectives for ITER related activities, Jour. Nuc. Mat. vol. 367-370, 1422-1431, 2007.
[33] Herrmann A, Interface Optimization of Tungsten Fiber-Reinforced Copper for Heat Sink Applications, PhD-thesis, IPP, TU-München, 2009.

2. Metal matrix composites

2.1. MMC categories

For engineering applications, metals and ceramics with different micro-structure, crystal-structure and bonding type represent two material classes with entirely different properties [1]. A metal with metallic bond shows good electrical / thermal conductivity and commonly ductility [2]. A ceramic with covalent bonding has a low electrical conductivity but stiff and brittle properties. Some applications require materials with properties of a metal combined with a ceramic. Metal matrix composites (MMC) are developed to combine both types of materials by reinforcing a ductile metal matrix with a high strength ceramic phase [3]. The macroscopic properties appear as a combination of the matrix with the reinforcement (rule of mixture). Some major improvements can be achieved like: increasing yield strength, increasing tensile strength, better high temperature creep resistance, improved fatigue strength, high thermal shock resistance, good corrosion resistance, increasing Young's modulus, decreasing thermal expansion. Relevant types of composites and their application are distinguished in Tab. 2.1. [1, 3, 4].

MMC	Reinforcement	Matrix	Embedded Phases	Production	Properties	Application
PRM	p: isotropic by particles	Mg, Al, Ti, Cu, Ag	Al2O3, TiB2, B4C, SiO2, TiC, WC, Diamond...	preform infiltration, powder metallurgical, stir casting	wear resistance, thermal properties, stiffness	brake caliper material, heat sink material
SFRM	s: random / planar oriented short fibers or w: whiskers	Mg, Al	Al2O3 Saffil® fibers, SiC whiskers	preform infiltration, powder metallurgical	elevated temperature strength	piston alloys, brake discs
CFRM	unaxial / multiaxial continuous fiber bundles	Mg, Al, Cu	carbon fibers, SiC fibers, Alumina fibers	preform infiltration, powder metallurgical	thermal properties, strength, stiffness	satellite antenna, heat sink material
MFRM	f: unaxial continuous monofilaments	Mg, Al, Ti, Cu	SiC fibers, W wires	diffusion bonding, coating	thermal properties, strength and stiffness at high temperatures	turbine blade, heat sink material
ICM	i: interpenetrating rigid single or multiphase sponge	Al, Cu	SiC+Si, CD+Si, sintered particles	preform infiltration, diffusion bonding,	wear resistance, thermally stable, stiff	brake discs, heat sink material

Table 2.1.: Types of metal matrix composites (MMC), examples of constituents and some engineering applications:

It is not only the kind of material used as reinforcement that is responsible for the macroscopic properties of the composite but also size and shape of the reinforcement. A classification of MMCs in fiber, short fiber and particle reinforced composites is shown in Fig. 2.1. The denomination of MMC according to [ASTM] indicates the constituents as follows:

matrix / reinforcement / vol.%, type - architecture

The shape of the reinforcement is indicated by the index quoted in Tab. 2.1. MMC considered here are linked accordingly in Fig. 2.1.

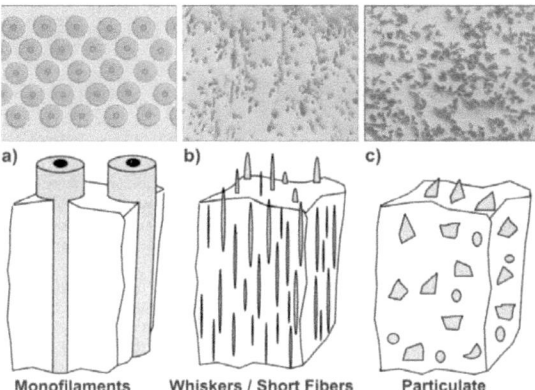

Fig. 2.1.: Metallographic cross sections (above) [4] and sketches (below) [3] of three composite groups: (a) Monofilament reinforced composite (MFRM); (b) short fiber reinforced composite (SFRM); particle reinforced composite (PRM) [1, 4].

The material properties of particle reinforced composites (PRM) such as stiffness, thermal conductivity and thermal expansion are isotropic on a macroscopic scale due to a spherical symmetry of their reinforcement architecture. Dense particle packing results in composites where high volume fractions of up to 60 vol.% (monomodal or polymodal particle sizes) and even higher volume fractions < 75 vol.% (polymodal only) can be realized. The higher the particle volume fraction rises the more their properties dominate [5, 6].

Fiber reinforced composites are an anisotropic reinforcement type. The thermo-mechanical properties are improved along fiber direction mainly. Monofilament reinforced composites (MFRM) represent a reinforcement type with uni- or bidirectional fiber orientations which are arranged first and then embedded into a matrix in a controlled direction (according to the macroscopic properties) by diffusion bonding [7, 8, 9]. The monofilaments are comparably large (Ø ~ 100 μm), and often extend over the whole component. Contrary short fiber reinforced materials, usually produced by infiltration of short fiber preforms (produced as such or chopped continuous fibers) are completely embedded in a matrix [10]. Random or planar orientations will result from their initial arrangement in the preform. The microscopic properties of ceramic fibers are isotropic longitudinal and transversal and influence the macroscopic thermo-mechanical properties due to the fiber arrangement [11]. If the reinforcement is interconnected, an interpenetrating composite will be achieved [12]. Particle or fiber preforms can be produced with ceramic binder or can be sintered, or a phase from the matrix forms bridges between the reinforcement as Si does from eutectic decomposition of an Al-Si or Ag-Si matrix.

2.2. Material properties

2.2.1. Thermal conductivity

For heat sink materials thermal conductivity is obviously one of the most important features [13]. The heat flux q through a material depends on the temperature difference ΔT, the path's geometry: area A, length l and the thermal conductivity λ (also TC) medium dependent as shown in equ. 2-1.

$$q = \frac{dQ}{dt} = \lambda \cdot \frac{A}{l} \cdot \Delta T \qquad (2\text{-}1)$$

In solid state, the heat flux can be established by phononic as well as by electronic conductance [2]. In metals electronic conductance dominates due to the metallic bond with valence electrons between the atoms. Good thermal conductivity "generally" goes with good electrical conductivity. In covalent bonded materials like ceramics or nonmetallic crystalline structure the phononic heat transfer is dominant due to a lack of valence electrons and a strong covalent bond.

In metal matrix composites the overall thermal conductivity is a combination of the conductance of the matrix and the reinforcement. The Maxwell approach [1] deals with spherical particles embedded in a matrix resulting as in equ. 2-2.

$$C = \frac{2 \cdot (1-f) + R \cdot (1+2f)}{(2+f) + R \cdot (1-f)} \qquad (2\text{-}2a)$$

with composite's conductivity relation C, reinforcement's conductivity relation R and volume fraction f.

$$(1-f) \cdot (R-1) = R \cdot C^{-n} - C^{1-n} \qquad (2\text{-}2b)$$

$$n = \frac{\frac{1}{X}}{\frac{1}{X} + \frac{1}{Y} + \frac{1}{Z}} \qquad (2\text{-}2c)$$

with shape factor n of the reinforcing ellipsoids given by the ellipsoids semiaxis X, Y and Z.

The conductivity of the composite $\lambda_c = \lambda_m \cdot C$ and the reinforcement is set in ratio C to the thermal conductivity of the matrix λ_m with $\lambda_m > \lambda_m \cdot C > \lambda_r$ and conductivity of the reinforcement $\lambda_r = \lambda_m \cdot R$. The heat flux through a composite is given by equ. 2-3:

$$q_i = -\lambda_{ij} \cdot T'_j \qquad (2\text{-}3)$$

with anisotropic thermal conductivity λ and transformation thermal gradient T.

and depends on the reinforcement shape. In long fiber reinforced composites the conductivity relation can be simplified in the two directions relative to fiber orientation in equ. 2-4.

$$q_{1c} = f \cdot q_{1r} + (1-f) \cdot q_{1m} \tag{2-4a}$$

$$q_{2c} = q_{2r} = q_{2m} = f \cdot T'_r + (1-f) \cdot T'_m \tag{2-4b}$$

with heat flux q, temperature T, reinforcement volume fraction f, suffix: m matrix, r fiber reinforcement, c composite, 1 longitudinal, 2 transversal.

The heat flux q is calculated longitudinally and transversally to fiber orientation, as shown in Fig. 2.2.

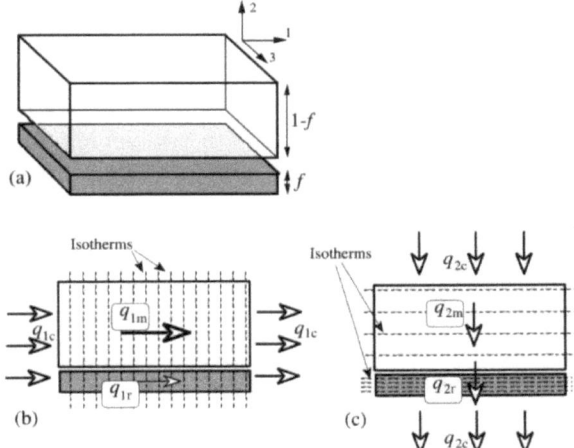

Fig 2.2.: Heat flux along and transversal to a fiber reinforced composite approximated by the slab model [1].

The effect of interface resistance was neglected in the calculations above but is of great importance in reality. In MMC the bonding quality is often inadequate, resulting in a higher interface resistance due to infiltration voids and cracks [14]. As well the heat transfer from one phase to the other will be hindered by different conduction types like electronic in the metal matrix and phononic in the ceramic reinforcement [2]. A realistic heat flux model is difficult to establish for complex structures like in particle or fiber reinforced composites, but this first approximation is usually assumed as sufficient.

2.2.2. Thermal expansion

In solids, the averaged interatomic spacing depends on superimposed attracting and repulsing forces between the atoms. The Lennard-Jones potential (equ. 2-5) [15] describes these overlapping potentials (Fig. 2.3).

$$V_{LJ} = -4\varepsilon \left[\left(\frac{R}{r} \right)^{12} - \left(\frac{R}{r} \right)^{6} \right] \quad (2\text{-}5)$$

with Lennard-Jones potential V_{LJ}, attracting force ε, particle distance r and finite distance to potential crossing the r-axis R.

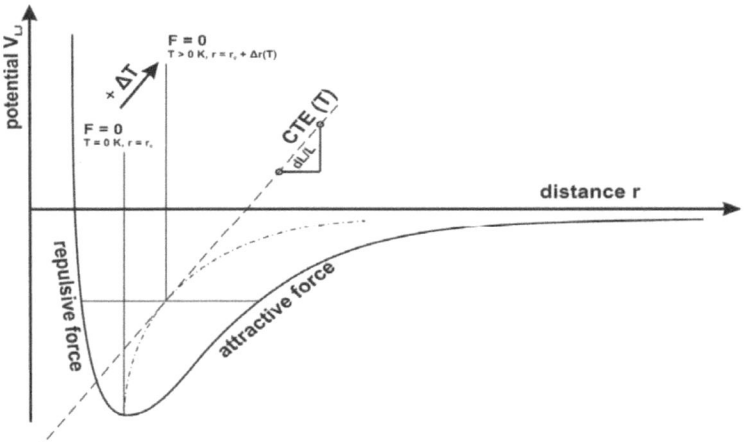

Fig. 2.3.: Lennard-Jones potential with superimposed attractive and repulsive contribution. The resulting potential well defines the averaged atomic distance as a function of energy if the wave function does not change with temperature.

The atoms get attracted by van der Waals or electromagnetic interaction, which are counteracted by short range Pauli repulsion of the electrons. An atom in its ground state (at T = 0 K) would take the lowest energy level on the bottom of the potential well with an interatomic distance r. With increasing temperature, the atom rises in the potential increasing the mean value of r. The increase of the atomic distance can be defined as volume or length change per unit increase of temperature (equ. 2-6).

$$\alpha_V = \frac{1}{V}\left(\frac{\partial V}{\partial T}\right) \qquad CTE = \frac{1}{L}\left(\frac{\partial L}{\partial T}\right) \qquad \frac{\Delta L_{21}}{L_1} = \frac{L_2(T_2) - L_1(T_1)}{L_1(T_1)} \quad (2\text{-}6)$$

with volumetric coefficient of thermal expansion α, linear coefficient of thermal expansion CTE and length L.

The instantaneous volumetric and linear coefficient of thermal expansion, α and CTE, can be calculated from the volume change V and the length change L, respectively. Composites are to be developed in which ceramic reinforcements hinder the metal matrix to expand or to contract, when T is changed. High micro stresses cause an averaging of the macroscopic CTE of a composite between matrix and reinforcement weighted by their volume fractions [1]. The thermo-elastic Turner model [16] allows the calculation of the CTE of a composite with CTE, Young's modulus E and volume fraction V of the constituents in equ. 2-7. It assumes the connectivity of the constituents, as IPC or CFRM (Tab. 2.1) in fiber direction.

$$CTE_c = \frac{CTE_r \cdot v_r \cdot E_r + CTE_m \cdot v_m \cdot E_m}{v_r \cdot E_r + v_m \cdot E_m} \qquad (2\text{-}7)$$

with linear coefficient of thermal expansion CTE, Young's modulus E, volume fraction v, suffix: c composite, r reinforcement, m matrix.

This fully elastic approach neglects matrix plastification, interface delamination and matrix voids [17]. In reality, high micro stresses in composites, produced by the big CTE mismatch during changing temperatures, are the main cause for irreversible thermal fatigue damage which will reduce the initially good thermal properties of the composite under service conditions and limit its long term stability [18].

2.3. Internal stresses

2.3.1. Mechanical load

In metal matrix composites strengthening is a consequence of load transfer from a soft matrix metal through an interface into a stiff ceramic phase [1, 3]. External load is distributed from the matrix metal through the interfaces into the stiff ceramic phase. Good interface bonding and high Young's modulus of the reinforcement are required for these composites. The macroscopic Young's modulus of the composite is combined by both constituents by using isostrain conditions as in equ. 2-8:

$$E_c = (1-f) \cdot E_m + f \cdot E_r \qquad (2\text{-}8)$$

with Young's modulus E, reinforcement volume fraction f, suffix: m matrix, c composite, r reinforcement.

The rule of linear mixture holds for IPC and CFRM within the elastic range. The elastic properties of the phases contribute weighted by their volume fraction. The stress distribution in a composite under external mechanical load is shown in Fig. 2.4.

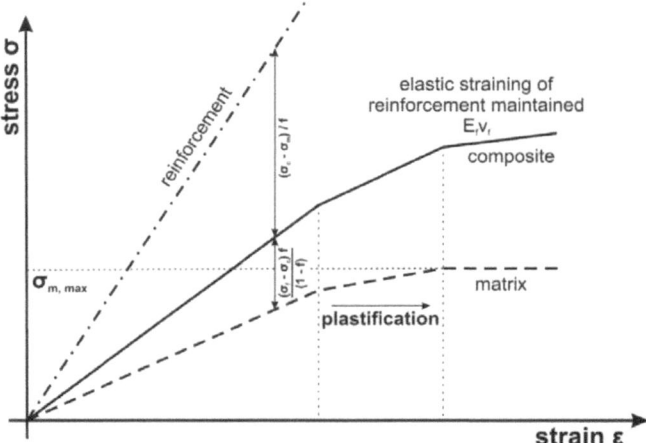

Fig. 2.4.: Stress strain diagram of a composite with a ductile matrix and stiff reinforcements assuming a persistant elastic range of the reinforcement without fracture.

The stiff reinforcement relieves load from the matrix reducing matrix strain. In the elastic region, the load distribution between matrix and reinforcement fulfills the rule of mixture according to the volume fractions. Surpassing the matrix yield strength produces irreversible deformation of the composite. Matrix plastification between the reinforcements causes strain hardening in the composite. Further increase of the external load results into fracture of the reinforcement and thus the composite. Unloading from the plastifying region before surpassing the composite's yield strength maintains macroscopic deformation that remains with residual stresses between both phases. In practice an initial splitting of the phase stresses caused by micro stresses originating from the production process, would be observed. These initial stresses are generated by matrix shrinking during cooling between stiff reinforcements due to their big CTE mismatch.

In fiber reinforced composites the basic composite mechanics are different due to a macroscopic heterogeneity, according to the fiber orientation. A first approach is the simplified slab model dealing with an equal strain condition into fiber direction only [3]. An equal stress condition is assumed transverse to fiber direction as shown in Fig. 2.5.

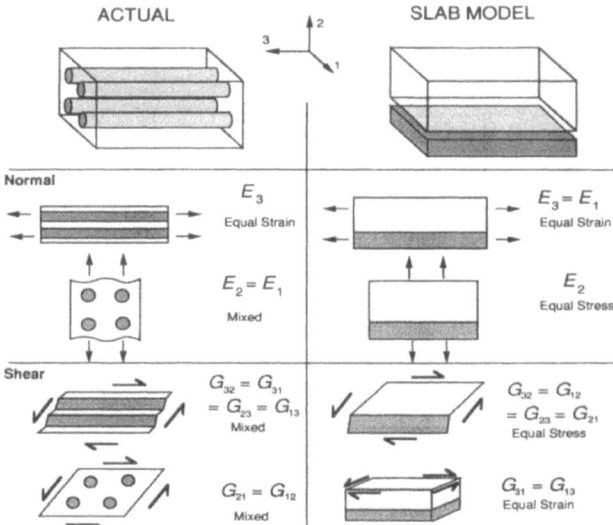

Fig. 2.5.: The slab model describing a unidirectional reinforced composite with equal strain situation longitudinal and equal strain transversal to fiber orientation [3]:

The same linear approach of the Young's modulus can be made as in equ. 2-8, neglecting any transversal (Hoop or radial stresses) or mixed contributions (Poisson's).

For particle reinforced composites the exactness of the model can be improved by introducing a particle shape factor. This shape factor includes a surface to volume relation of the particles, relevant for the overall Young's modulus in the Tsai Halpin model [3] (equ. 2-9):

$$E_c = \frac{E_m (1 + 2 \cdot S \cdot q \cdot f_p)}{1 - q \cdot f_p} \qquad (2\text{-}9)$$

with Young's modulus E, shape factor S, suffix: m matrix, p particle.

and geometry factor q in equ. 2-10:

$$q = \frac{\left(\dfrac{E_p}{E_m}\right) - 1}{\left(\dfrac{E_p}{E_m}\right) + 2 \cdot S} \qquad (2\text{-}10)$$

with Young's modulus E, shape factor S, suffix: m matrix, p particle.

2.3.2. Thermal load

Thermal stresses are produced by changing temperatures in MMC with a big CTE mismatch. High micro stresses near the matrix-reinforcement interfaces will lead to thermal fatigue damage such as matrix damage and delamination which are therefore of main interest [3]. During thermal cycling, these stresses will oscillate between tension and compression Fig. 2.6.

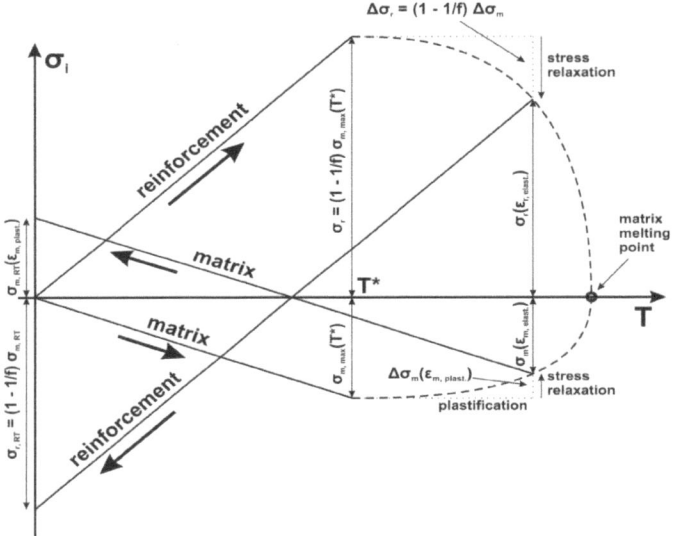

Fig. 2.6.: Thermal micro stresses in a composite with a soft ductile matrix with stiff reinforcement. CTE mismatch stresses are weighted indirectly proportional to the volume fraction of the corresponding phase. The ductile phase gets plastified according to its temperature dependent yield strength. $\sigma_i = 0$ is assumed as RT starting condition.

This example shows a matrix with a bigger CTE than that of the reinforcement and with a lower E_m < E_c, as expected for MMCs. The matrix expansion is suppressed by the stiff reinforcements with low CTE (compared to the matrix) during heating. This compression is compensated by tensile stresses in the reinforcement fulfilling the equilibrium condition. The stress amplitudes of both are proportional to their volume fractions so that an average over the whole volume of the composite will be zero. At elevated temperatures, plastification of the softer phase produces stress hysteresis during thermal cycling. Residual stresses remain after cooling to room temperature by matrix shrinking with stress inversion in high-temperature-relaxed regions. A similar stress level can be expected as well right after production (infiltration/consolidation at high temperatures) in any composite with a big CTE mismatch. The averaged misfit strain induced by changing temperatures can be calculated with equ. 2-11.

$$\varepsilon^T = \int_{T_1}^{T_2} (\alpha_I - \alpha_M) dT \approx (\alpha_I - \alpha_M) \cdot \Delta T \qquad (2\text{-}11)$$

with strain ε, volumetric coefficient of thermal expansion α, thermal gradient ΔT, suffix: I inclusion, M matrix.

The simple first approximation [3] assumes an isotropic geometry with isotropic expansivities of the inclusion. The straining is assumed relative to the matrix in the inclusion only. In reality, the misfit strain depends on shape effects of the inclusion in order to matrix deformation inverse to the inclusion and on plastification near the interfaces. To improve the understanding of misfit strains in composites, the Eshelby model is introduced.

2.3.3. Eshelby model [19, 20]

The basic idea of this model is to cut the reinforcements out of a matrix, deform them freely according to the imposed load and to reassemble them again (Fig. 2.7). At first, the case of mechanical load on the misfit stresses is discussed, which can be calculated using equ. 2-11.

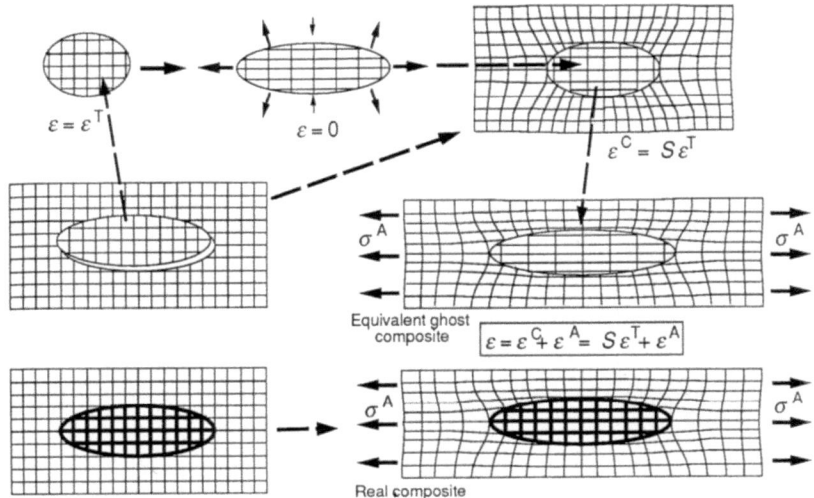

Fig 2.7.: Eshelby model of a reinforced composite with externally applied load [3]. The transformation strained ghost inclusion replaces the uniformly strained real inclusion.

The external tension causes deformation of the composite, by inducing internal stresses between matrix and reinforcements. These stresses can be calculated with the difference between the freely deformed and constrained inclusion. The internal stresses result from the stiffness tensor of the material and equ. 2-12.

$$\sigma_I = C_M \cdot (\varepsilon^C - \varepsilon^T) \qquad (2\text{-}12)$$

with stress σ, stiffness tensor C, strain ε, suffix: C constrained, T transformation, M material.

By introducing the Eshelby 'S' tensor, the aspect ratio and the Poisson's ratio of the reinforcement are included (equ. 2-13).

$$\varepsilon^C = S \cdot \varepsilon^T \qquad (2\text{-}13)$$

The Eshelby S tensor contains the elastic energy (shape dependent) in the constrained region near an inclusion and indicates the relation from strain to the transformation strain field. The stress free shape misfit is thus given by equ. 2-14.

$$\sigma_I = C_M \cdot (S - I) \cdot \varepsilon^T \qquad (2\text{-}14)$$

with stress σ, stiffness tensor C, Eshelby tensor S, identity matrix I suffix: I internal, M material, T transformation.

Also thermal misfit stresses are calculated with the Eshelby model by taking an inclusion with a lower CTE out of the surrounding matrix (Fig. 2.8).

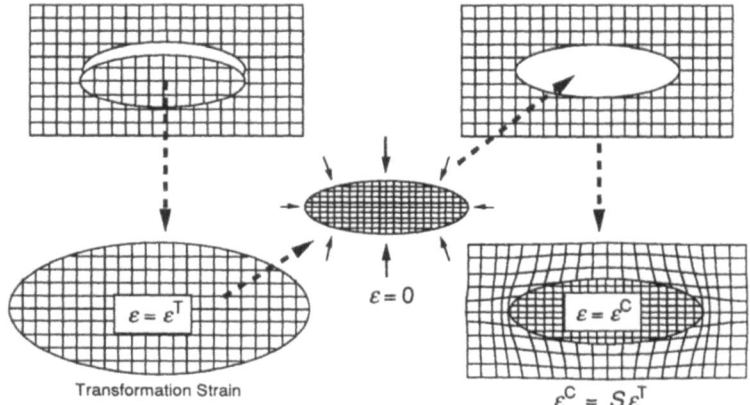

Fig. 2.8.: Eshelby model for thermal misfit strain in a composite [3]. The natural shape misfit between the hole and the inclusion results in a constrained inclusion and surrounding matrix.

Both, inclusion and matrix are cooled down. The inclusion will not shrink as much as the matrix, which will induce stresses after reassembling the big particle in a too small hole in the matrix. In this case, the stresses are given the stiffness tensor and the different thermal strain (equ. 2-15).

$$\sigma_I = C_I \cdot (\varepsilon^C - \varepsilon^{T*}) \tag{2-15}$$

with stress σ, stiffness tensor C, strain ε, suffix: C constrained, T transformation.

Compared to the linear external load condition, the Eshelby 'S' tensor relation equ. 2-13 will not be further equal but result in equ. 2-16.

$$\varepsilon^C > S \cdot \varepsilon^{T*} \tag{2-16}$$

An overall strain value will be generated in the inclusion as well as in the matrix in order to calculate the strain in the inclusion. The matrix strain has to be implemented in the stress equilibrium condition and a calculated CTE, depending on elastic moduli and volume fractions of both phases.

The CTE of the composite calculated with the elastic Turner model [16] is used as starting point of strain, which allows linear stress calculation with equ. 2-17 [5].

$$\sigma_m = E_m \left(CTE_c - CTE_m \right) \Delta T \tag{2-17a}$$

$$\sigma_r = E_r \left(CTE_c - CTE_r \right) \Delta T \tag{2-17b}$$

with stress σ, linear coefficient of thermal expansion CTE, Young's modulus E, temperature difference ΔT and suffices: r reinforcement, m matrix and c composite.

The stresses are calculated in the components each relative to an overall composite thermal expansion. The different expansions $CTE_r < CTE_c < CTE_m$ at the same temperature changes result in internal stresses fulfilling the equilibrium condition in the composite volume averaged over all phases.

References

[1] Mortensen A, Kelly A, Zweben C, Comprehensive Composite Materials, vol. 3, Metal Matrix Composites, 541-547, Oxford 2000.
[2] Harrison WA, Electronic Structure and the Properties of Solids, W. H. Freeman and Company, San Francisco, 1980.
[3] Clyne TW, Withers PJ, An Introduction to Metal Matrix Composites, Cambridge University Press, Trumpington street, Cambridge CB2 1RP, 1993.
[4] http://mmc-assess.tuwien.ac.at, May 2011.
[5] Huber T, Degischer HP, Lefranc G, Schmitt T, Thermal expansion studies on aluminium-matrix composites with different reinforcement architecture of SiC particles, Comp. Sci. Tech. vol. 66, 2206-17, 2006.
[6] Arpon R, Molina JM, Saravanan RA, Carcia-Cordovilla C, Louis E, Narciso J, Thermal expansion behavior of aluminium/SiC composites with bimodal particle distributions, Acta Mat. vol. 51, 3145-3156, 2003.
[7] Leyens C, Kocian F, Hausmann J, Kaysser WA, Materials and design concepts for high performance compressor components, Aero. Sci. Tech. vol. 7, 201-210, 2003.
[8] Brendel A, Popescu C, Köck T, Bolt H, Promising composite heat sink material for divertor of future fusion reactors, Jour. Nuc. Mat. 367-370, 2007.
[9] Peters P, Hemptenmacher J, Schurmann H, The fiber/matrix interface and its influence on mechanical and physical properties of Cu-MMC, Comp. Sci. Tech. vol. 70, iss. 9, 1321-1329, 2010.
[10] Dlouhy A, Merk N, Eggeler G, A microstructural study of creep in short fiber reinforced aluminium alloys, Acta. Metall. vol. 41, iss. 11, 3245-3256, 1993.
[11] Requena G, Fiedler G, Seiser B, Degischer P, Di Michiel M, Buslaps T, 3D-Quantification of the distribution of continuous fibers in unidirectionally reinforced composites, Comp. part A, vol. 40, iss. 2, 152-163, 2009.
[12] Lefranc G, Degischer HP, Sommer HK, Mitic G, In: Massard T, editor. ICCM12 proceedings, Paris, electronic support, 1999.
[13] Nogales S, Böhm HJ, Modeling of the thermal conductivity and thermomechanical behavior of diamond reinforced composites, vol. 46, iss. 6, 606-619, 2008.
[14] Battabyal M, Beffort O, Kleiner S, Vaucher S, Rohr L, Heat transport across the metal-daimond interface. Diamond and related Materials, 17, 2008.
[15] Ashcroft MW, Mermin ND, Solid State Physics, South Melbourne, Victoria: Books/Cole, 2005.
[16] Turner P, Thermal-expansion stresses in reinforced plastics, J Re Nat Bu Stan, 36, 239-50, 1946.
[17] Stinchcomb WW, Ashbaugh NE, Composites Materials: Fatigue and Fracture, vol. 4, ASTM, 1916 Race Street, Philadelphia, PA 19103, 1993.
[18] Shapery RA, Thermal expansion coefficients of composite materials based on energy principles, Jour. of composite materials 2, 380-404, 1968.
[19] Eshelby JD, The Determination of the Elastic Field of an Ellipsoidal Inclusion and Related Problems, Proc. Roy. Soc. (London) A241, 376-396, 1957.
[20] Eshelby JD, The elastic field outside an ellipsoidal inclusion, Proc. Roy. Soc. (London) A252, 561-569, 1959.

3. Investigated MMC

3.1. Particle reinforced metals (PRM)

In particle reinforced composites a ductile matrix gets reinforced by stiff particles to change the thermo-mechanical properties of the composite adequately. Heat sink applications require the combination of high TC with low CTE.

3.1.1. Al-SiC

In power electronics, aluminum alloys reinforced with ceramic SiC particles emerge as successful composite [1]. These metal matrix composites combine the good thermal conductivity of an Al-matrix with the low thermal expansion of ceramic SiC-particles (Fig. 3.1).

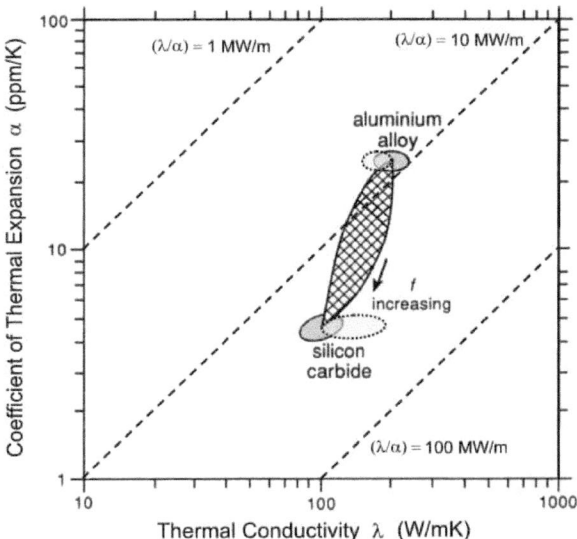

Fig. 3.1.: CTE vs. TC showing the perspectives in the thermal properties achievable, combining an Al matrix with SiC particles [2].

Volume fractions up to 70 vol.% at different size distributions of the SiC particles reduce the CTE down to 6 – 10 ppm/K according to the rule of mixture [3]. Fig. 3.2 shows an AlSi7 matrix reinforced with ~ 60 vol.% of SiC particles bimodal (Ø ~ 5 and 50 μm).

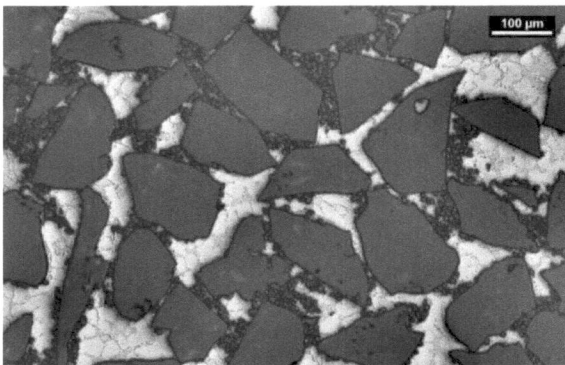

Fig. 3.2 Metallography of AlSi7/SiC/60p with bimodal particles (dark) embedded in an AlSi7 matrix (bright) with eutectic regions (grey) in between [4].

SiC particles are used due to their good bonding with Al alloys. The low thermal conducticity of SiC (TC_{SiC} ~ 140 W/mK) combined with pure Al (TC_{Al} ~ 240 W/mK) would reduce the overall conductivity of the composite to TC_{MMC} ~ 180 W/mK, but also decreases the overall thermal expansion from the matrix (CTE_{Al} ~ 25 ppm/K) down to CTE_{MMC} ~ 6 – 10 ppm/K of the composite, dependent on the SiC (CTE_{SiC} ~ 4 ppm/K) particle volume fraction [2]. Actually composites with suitable thermo-mechanical properties for heat sink materials in power electronic applications are realized with AlSiC. Several Al matrix alloys are used with different benefits under operation conditions. Pure Al offers the best TC compared to AlSi alloys. Si content in the matrix increases bonding strength with the particles by a controlled reactivity with SiC and avoids their dissolution. As well < 1 at.% of Mg is added in commercial AlSi7Mg alloys for precipitation hardening of the α-Al between the Al-Si eutectic (intrinsic matrix reinforcement) [5]. Both Si and Mg impurities reduce the thermal conductivity but increase matrix and bonding strength. The mechanical strength of the components in the composite is important to withstand thermal mismatch stresses under operation conditions and to reduce thermal fatigue damage during changing temperatures. Si in the matrix does not only improve bonding / matrix strength but also changes the reinforcement architecture reducing the fatigue damage propagation as discussed later. Moreover, a certain amount of voids is formed after production by liquid metal infiltration [2]. The big CTE mismatch (ΔCTE ~ 20 ppm/K) between the Al matrix and the densely packed SiC particles generates voids < 2 vol.% by a shrinking matrix during cooling from 650 °C [3]. These voids arranged at the interfaces would reduce the TC of the composite, and their volume fraction has to be minimized as much as possible.

3.1.2. Diamond reinforced metals

Carbon diamond (CD) is a promising candidate as reinforcement for the next generation heat sink composites [6, 7]. Diamond does not only offer a lower thermal expansion (CTE_{CD} ~ 1 ppm/K) compared to SiC (CTE_{SiC} ~ 4 ppm/K), but also represents the material with the highest thermal conductivity (TC_{CD} ~ 1000 - 2000 W/mK) known (Fig. 3.3).

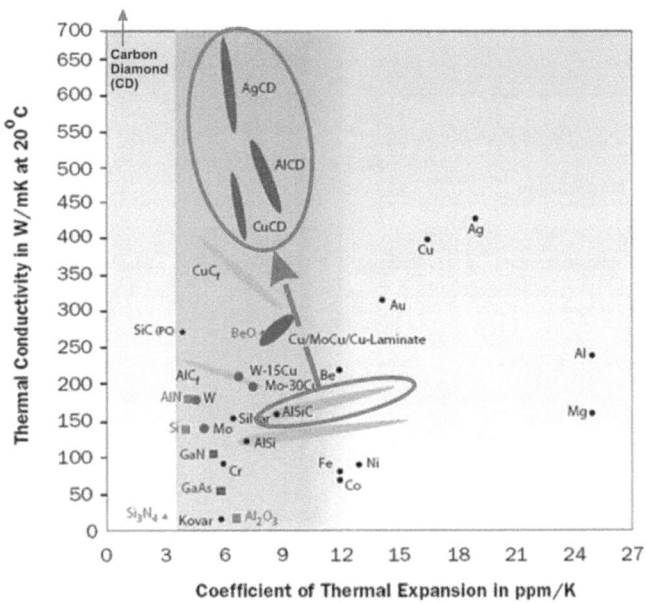

Fig. 3.3.: The TC vs. CTE of some common materials showing the expected archievement by diamond reinforced metals compared to AlSiC [8].

A metal matrix reinforced with diamond particles does not only benefit from the lower CTE but also display a significantly higher TC. Several high conducting matrix alloys, such as Al, Cu and Ag, are tested concerning their capability to be reinforced by diamonds [4, 9, 10, 11]. The reactivity of diamond with the metal during liquid metal infiltration is important for wettability and bonding strength. Al represents a capable matrix material due to its carbide forming ability [6]. Additional Si content is applied inspired by the AlSi7-SiC system to increase bonding and matrix strength. Higher thermal stresses can be expected at changing temperatures between diamond and aluminum due to a higher CTE mismatch (ΔCTE ~ 24 ppm/K). Heat treatments are performed to increase bonding strength, supporting carbide formation at the interfaces [7]. Copper matrix alloys offer a better thermal conductivity and a lower CTE mismatch (ΔCTE ~ 16 ppm/K), but without reactivity of liquid copper with carbon compared to an Al matrix. CuB alloys are developed with better adhesion of the diamond particle interfaces to improve bonding strength capable to withstand the temperature changes as expected under service conditions [12]. For some advanced thermal management solutions in space application, such as satellite laser optics, where the raw material price of a component is secondary, Ag matrix alloys are interesting. Ag does not only contribute a higher thermal conductivity compared to other matrix metals but also offers good reactivity when alloying with Si. These high performance heat sink materials of diamond reinforced AgSiX matrix alloys are tested concerning their thermal conductivity, interface quality and capability for applications comparable to the lower cost segment of diamond reinforced AlSiX matrices in power electronics. Fig. 3.4 shows fracture surfaces of AlCD, CuCD and AgCD composites.

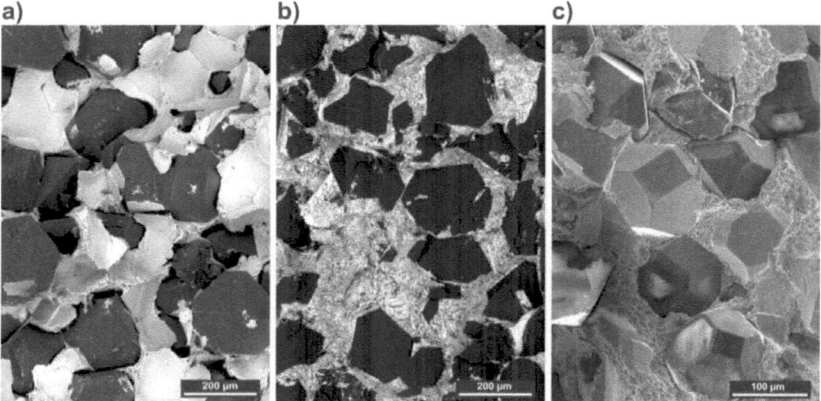

Fig. 3.4.: SEM images of fracture surfaces of a diamond reinforced (a) Al (by EMPA), (b) AgSi11 (by Plansee) and (c) CuB1 (by IFAM) matrix alloy.

Diamond particles are promising to replace the SiC particles as reinforcement of a high conducting metal matrix. With high volume fractions (60 vol.%) of diamond mono crystals, composites with superior thermal conductivity (TC_{AlCD} > 500 W/mK) can be realized (TC_{AlSiC} ~ 300 W/mK) [8, 13]. The interface bonding quality of the plane crystallographic surfaces is important for good thermal contact between the matrix metal and the particle and to withstand CTE mismatch stresses. Particle sizes are tested concerning their thermal conductivity and mechanical stability (Ø ~ 25 – 200 µm). A big particle reduces the number of interfaces to be passed by the heat flux, and small particles reduce hoop stresses and interface area (mismatch stresses) [14, 15]. The main problem of heat dissipation through a matrix-particle interface is caused by two types of thermal conductors. The matrix metal, an electronic conductor, has to pass heat to the diamond, a phononic conductor. Big particle sizes improve thermal conductivity due to a lower amount of interfaces. Highest particle volume fractions of 60 vol.% monomodal with particle sizes of Ø ~ 100 µm are promising.

3.1.3. PRM production process

High volume fraction particle reinforced metals are usually produced by melt infiltration or powder metallurgy [14]. The particle reinforced metal matrix composites described in this work were produced by pressure driven infiltration of a particle preform with liquid matrix metal. External pressure is applied during infiltration in order to reduce defects like voids and to overcome poorly wetted interfaces. Two different production types for particle reinforced composites were used.

Gas pressure infiltration

The process of gas pressure infiltration can be divided into several steps as shown in Fig. 3.5. A pressure vessel containing a particle preform and the matrix metal are evacuated. Then the chamber is heated up to melting temperature of the matrix alloy which lies on top of the preform, incompletely wetting the particle powder. Inert gas injection (Argon) into the chamber presses (< 10 MPa) the melt on top into the evacuated powder. During directional cooling the metal matrix solidifies under external hydrostatic compression improving the infiltration quality.

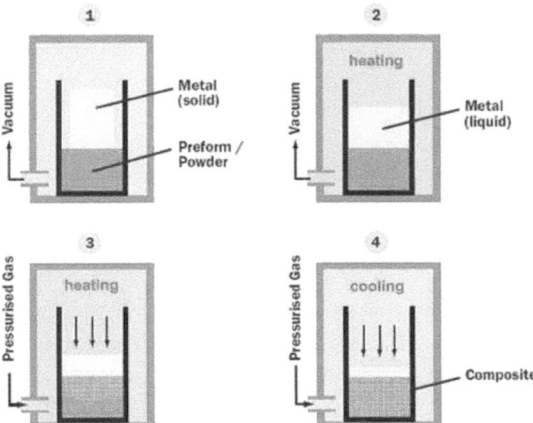

Fig. 3.5.: The gas pressure infiltration process divided into its four stages [8].

The remaining melt above the particle powder feeds matrix shrinking between the particles. With this production route shaped products with high degree of complexity can be manufactured.

Squeeze casting

During squeeze casting similar procedures can be distinguished comparable to gas pressure infiltration [14, 16] (Fig. 3. 6).

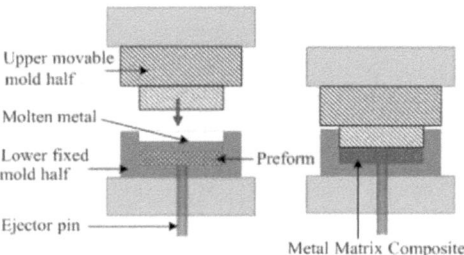

Fig. 3.6.: The squeeze casting process [16].

A particle containing preform is preheated and filled with the molten metal matrix. A pressure piston is put on top pressing the liquid metal into the preform. Pressure ranging from 10 to 100 MPa is maintained during solidification achieving low initial void volume fractions. The shape of the component produced by squeeze casting is limited. In some cases, heat treatments of the composites after infiltration under pressure are realized in order to relax residual stresses and to increase thermally induced interface reactivity between the particles and the matrix metal.

3.1.4. Particles

SiC particles used for the AlSiC composites are produced by sintering or CVD processes and fragmentation [17]. SiC is similar to diamond, because Si is located in the same main group as C, only different in atomic radius. The polytypic crystal structure exists in several phases [18]. The cubic β-SiC with its Zinkblende structure is similar to diamond and exists beside hexagonal α-SiC phases which are always dominant. SiC is used as reinforcement of heat sink materials due to its good thermal conductivity ($TC_{SiC} \sim 140$ W/mK) and low thermal expansion ($CTE_{SiC} \sim 4$ ppm/K). The reactivity with Al matrix alloys allows high bonding strength feasible for application under cycling thermal load. Al-Si matrix alloys have to be used to avoid dissolution of SiC and formation of Al_3C_4.

Diamond particles are produced commercially by HPHT (High Pressure High Temperature) processes grown from graphite in a metal solvent [19]. Usually FeCo or FeNi molten catalysts are used to grow diamonds at ~ 1500 K under a pressure of ~ 5 GPa. The basic idea of growing diamonds is the different solubility of graphite and diamond in a metal inducing diamond crystallization and slow growth. Diamond particles are promising to replace SiC as reinforcement for future heat sink applications due to their superior thermal properties. The reactivity of diamond particles with molten metals encourages using them in liquid metal infiltrated preforms. In order to increase bonding strength interface treatments like coatings or roughening by etching are applied to the plane crystallographic surfaces of the diamond particles before infiltration [20].

3.2. Monofilament reinforced materials (MFRM)

Monofilament reinforced copper composites are developed as high temperature heat sink materials [21]. The highly thermal conductive Cu matrix (TC_{Cu} ~ 400 W/mK) gets reinforced with low CTE fibers reducing the thermal expansion in fiber direction. In elongated divertor elements, a MFRM interlayer is proposed in order to reduce thermal mismatch stresses between W and CuCrZr (ΔCTE ~ 11 ppm/K) as shown in Fig. 1.9. The stiff fibers reduce the CTE of the matrix into fiber direction with thermal stresses distributed in all three dimensions of the composite's volume. Transversal deformability of the ductile matrix will not damage the interfaces in CuCrZr/W/Xm. This may reduce the micro stresses in the composite during changing temperatures [22]. MFRM are favored for fusion reactor applications due to their capability of withstanding radiation embrittlement [23]. Stress induced crack formation and growth is stopped at the fiber matrix interfaces [24] resulting in an improved toughness under operation conditions. The long term stability of components in fusion reactor environments is improved by MFRM significantly. Two different composite types are developed. SiC-fiber reinforcement competes with W-wire reinforcement for application as future heat sink material for the pulsed operated Tokamak fusion reactor system. Interface designs like fiber coating, etching and grading are applied in order to improve bonding strength as described in the following passage.

3.2.1. SiC fiber reinforced copper

SiC ceramic fibers are used as reinforcement due to their low CTE, high stiffness, yield strength and creep resistance at high temperatures. The CTE of the ductile Cu matrix gets reduced by the stiff SiC monofilaments to an overall CTE_{CuSiC} ~ 17 ppm/K of the composite into fiber direction [25]. Commercially available, SCS6 and SCS0 SiC fibers [26] with a diameter of ~ 140 µm are embedded by coating a Cu matrix alloy and HIPing [27]. A cross section through Cu/SiC/20m, pure copper reinforced with ~ 20 vol.% of SiC monofilaments is shown in Fig. 3.7.

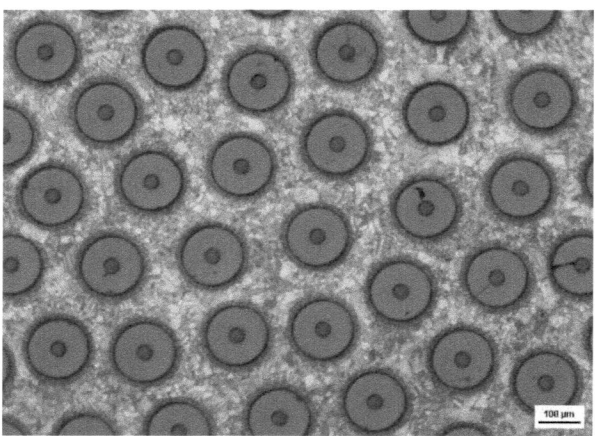

Fig. 3.7.: Metallography of a cross section through a Cu/SiC/30m (30 vol.% monofilaments) MFRM. Dark SiC fibers (Ø ~ 140 µm) embedded in a Cu matrix with visible grain structure [28].

The good thermal conductivity is mainly established by the high conducting matrix (TC_{Cu} ~ 400 W/mK), even reduced by the SiC fibers (TC_{SiC} ~ 16 W/mK, nano crystalline), but still sufficient for application (TC_{CuSiC} ~ 250 W/mK, dependent on fiber volume fraction). Heat dissipates predominantly in transversal direction between the fibers through the matrix from a tungsten plate into the CuCrZr heat sink (Fig. 1.9) [29]. Pure copper with higher thermal conductivity as well as precipitation hardened CuCr1Zr matrices are tested. The matrix with increasing ductility at high temperatures (dissolution of the precipitates ~ 450 °C) gets plastically deformed during heating accommodating the high thermal mismatch stresses. CuCr1Zr is introduced to withstand the tensile stresses in the matrix created during cooling. Ti coating of SCS6 SiC fibers improve bonding strength at the SiC-Cu interfaces. The well-known Ti-SiC MFRMs [30], successfully tested as turbine engine material for airplane application, inspired the development of a Ti interlayer (Ø ~ 200 nm) [31] of high reactivity. Some more complex TiTaC [32] coatings developed by IPP on

SCS0 SiC fibers were tested concerning bonding quality with different interface situations (SCS0 compared to SCS6). The tested fiber volume fraction varying between 10 and 30 vol.% has to be balanced between TC and σ_i (internal stress). A low volume fraction increases the thermal conductivity in the Cu matrix between the fibers, and a high volume fraction produces a more homogeneous stress distribution within the matrix. A thermal conductivity sufficient for application and internal stresses, not surpassing the matrix yield- / interface bonding-strength, has to be achieved.

3.2.2. W wire reinforced copper

The W-Cu system is developed parallel to Cu-SiC, using W wires as reinforcement with advantages in thermal conductivity ($TC_W \sim 170$ W/mK, $CTE_W \sim 5$ ppm/K), ductility and reactivity with copper [22]. Composites with suitable thermal properties ($TC_{CuW} \sim 300$ W/mK, $CTE_{CuW} \sim 17$ ppm/K) are realized. The handling during manufacturing the ductile W wires is less complicated than for the brittle SiC fibers. Good fiber-matrix bonding strength is achieved even without any interface pretreatment due to the reacting W-Cu interface [33]. W monofilament reinforced CuCr1Zr-MFRM with a wire volume fraction of 30 vol.% is shown in Fig. 3.8.

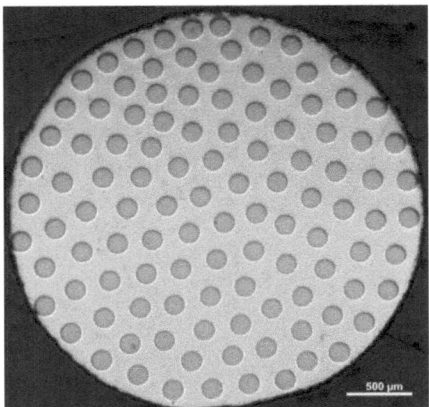

Fig. 3.8.: Cross section through a CuCrZr/W/30m tensile test sample (DLR) [27] as used for the thermal cycle tests during in situ neutron diffraction.

Additional surface roughening by etching and graded interface designs are applied to achieve highest bonding strength characteristics (IPP) [34]. The W wire acts as reinforcement similar to SiC by reducing the CTE of the composite in wire direction to an intermediate value between Cu and W suitable as interlayer material between the W and CuCrZr component of the divertor. Volume fractions of 20 to 50 vol.% were investigated. Comparable to the Cu/SiC system, the wire volume fractions in Cu/W-MFRM have to be balanced in order to achieve a good TC combined with low interfacial shear stresses under operation conditions.

3.3. MFRM production process

A lot of production or synthesis techniques are available for fiber reinforced metal matrix composites. Most common is liquid infiltration, comparable to the processing of particle reinforced composites, where the fibers arranged in a preform are infiltrated by a matrix melt. Or fibers between matrix foils are compacted by diffusion bonding at high temperatures [27, 30]. The production route of the composites investigated in this work was the patented compaction of matrix coated fibers developed by DLR. In this process three steps can be distinguished as illustrated in Fig. 3.9.

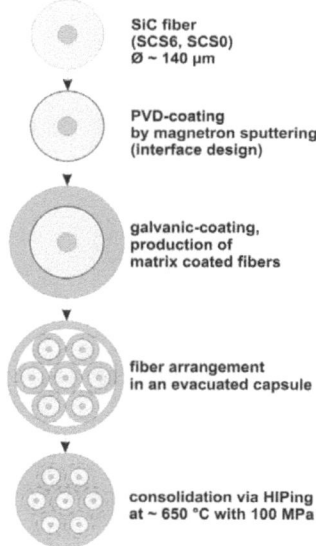

Fig. 3.9.: MFRM production process. PVD coated fibers are coated further galvanically by the matrix and consolidated via HIPing [31].

The first step is PVD (Physical Vapor Deposition) coating of the raw fibers [34] when interlayers are applied. The sputtering deposition process is also called magnetron sputtering [27]. The fibers are arranged in a vacuum chamber. Plasma is supplied by an inert gas ionized through a glow discharge. These ions are accelerated towards a cathode (target), evaporating atoms from the target material into the chamber. The gas atoms deposit on the substrate dependent on gas pressure and bias (negative voltage between substrate and base). The magnetron setup is used to catch free electrons in front of the target due to an additional magnetic field improving the sputtering process.

The second step is coating of the monofilaments by the matrix, that can be executed as well by magnetron sputtering or by galvanizing (electro chemical coating) of the fibers with matrix material [28]. The fibers (cathode) are positioned in a $CuSO_4$ liquid with a conducting Cu anode. A constant current through the electrolyte deposits the Cu ions onto the fibers. With increasing time a matrix coating with increasing thickness can be obtained, effecting the volume fraction of the consolidated composite (Fig. 3.10).

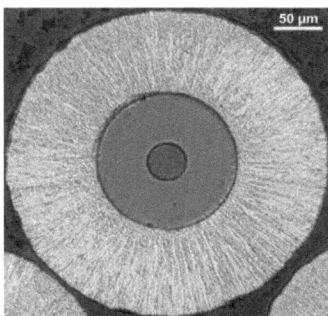

Fig 3.10.: LOM image through a cross section of a SiC fiber galvanically coated by matrix before consolidation [29].

After galvanizing of the fibers, additional heat treatments at ~ 550 °C with low heating rates ~ 20 K/min reduce voids and the hydrogen content in the matrix [29]. In case of reactive Ti interlayers, carbide formation is induced above 350 °C. In order to reduce damaging oxide on the copper layer, etching is performed immediately before consolidation.

Finally the matrix coated fibers are arranged in a matrix capsule (Fig. 3.11) which is evacuated and sealed by laser welding [28]. The fibers in the capsules are finally consolidated by hot isostatic pressing (HIP) at a temperature of 650 °C with a pressure of 100 MPa. The resulting unidirectional monofilament reinforced copper MMC has parallel arranged fibers well embedded in diffusion bonded matrix regions.

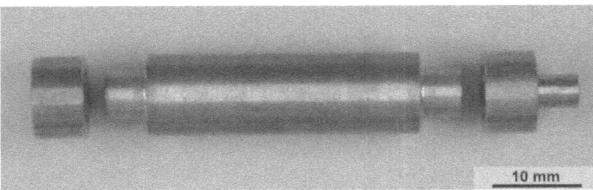

Fig. 3.11.: The capsule (preform) in which the fibers are arranged before evacuation and laser welding [28].

3.4. Monofilaments

Two types of SiC fibers have been used as reinforcement, the SCS0 and SCS6. The SCS6 fiber was developed for titanium and ceramic matrix materials for high temperature applications [26]. The SCS6 fiber consists of several shells as illustrated in Fig. 3.12.

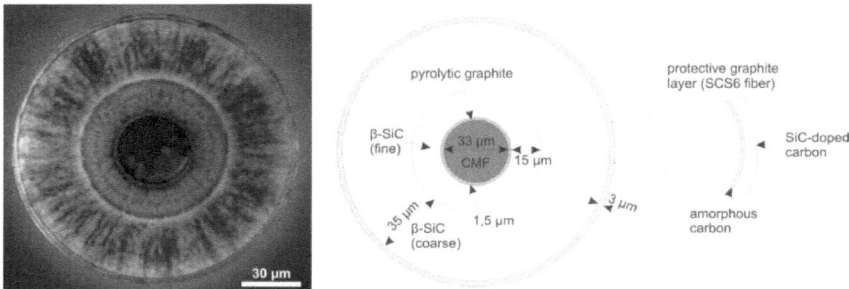

Fig 3.12.: A SCS6 fiber illustrated by a LOM image of its cross section (left) with its several layers distinguished (right) [26, 28].

The SiC monofilament with a diameter of ~ 140 μm is grown around a carbon filament substrate (CMF) which is arranged in the center, coated by pyrolytic carbon with 1.5 μm in thickness. On this graphite core a β-SiC layer is deposited by CVD (Chemical Vapor Deposition) coating. The inner region of ~ 15 μm has a nano grain structure (∅ ~ 1 – 30 nm) surrounded by an outer β-SiC region of ~ 35 μm with larger grains. The SCS6 fiber is protected by 3 carbon rich layers together 3 μm in thickness. The outer carbon layer acts as protection during handling and diffusion barrier for Si (to avoid formation of Ti-silicides) in Ti-matrix. This graphite layer of the SiC fiber also accumulates damage within the interface achieving a directed pullout of the fiber from the matrix gaining an improved damage tolerance of the composite. The as received SCS6 fibers are coated by magnetron sputtering. Thus selected interfaces can be applied to the SiC fiber before galvanic matrix coating and consolidation. In order to improve the reactivity of the interface between SiC fiber and Cu matrix, a highly reactive Ti interlayer of 200 nm is applied (PVD) [31]. Finally a thin (~ 500 nm) copper layer deposited on top, which protects the titanium against oxidation and improves bonding between Ti and the galvanic copper matrix in the following step (Fig. 3.13) [28].

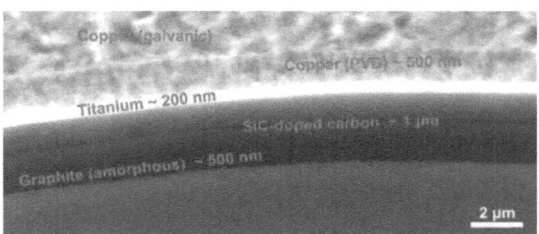

Fig. 3.13.: A closer electron microscopic view on the SiC-Cu interlayer of a Ti coated SCS6 fiber [31].

The SCS6 fibers without a protective graphite surface layer are coated with a Ti-TaC [32] interlayer system to achieve bonding between SiC and the Cu matrix. First a 40 nm tantalum interlayer is deposited onto the fibers, which are heat treated afterwards, activating a Ta-SiC reaction. A layer with combination of TaC and TiC is covered by a layer of TiC and then by a Ti layer. The final PVD coated Cu layer acts similarly as the SCS6 fibers. The complete TiTaC interlayer system of the SCS0 fibers gains a total thickness of ~ 570 nm.

W monofilaments are produced by a hot fiber drawing process at temperatures between 200 and 500 °C, where tungsten gets ductile and therefore deformable [35]. As received from production, the W monofilaments have to be ultrasonically and chemically cleaned before used. Then further surface treatments by etching or coatings are applied. Monofilaments without interface treatment get directly galvanized by the matrix, a PVD matrix pre-coating step [33] is integrated, similar to the finalizing Cu coating of the SiC fibers in order to improve the Cu-W adhesion. For improving bonding strength by interfacial form closure, an etching process is applied to the W wires which results in a surface roughening before galvanizing. Some more complex graded interface designs are realized by deposition (magnetron sputtering) of varying Cu and W concentrations. Goal is a radially graded interface with continuously growing Cu content from the fiber surface through a ~ 500 nm thick inter-layer into the Cu matrix.

3.5. Investigated samples

Several metal matrix composites for heat sink applications were investigated in this work. Representative types of fiber as well as particle reinforced systems have been delivered by our project partners. The composite samples were produced suitable in size and shape for the in situ diffraction and tomography experiments. Different interface designs as well as matrix alloys were tested in the same way in order to deliver comparable results. The material to show the capability of diffraction in combination with tomography as useful method was AlSi7Mg/SiC/70p from Electrovac Austria. The well-known material was tested in order to be able to answer open questions from previous work on some anomalous macroscopic thermal properties (CTE) [3]. This very first and successful campaign inspired the experimental one to follow [36]. Different types of AlSiC and AlCD were investigated comparably. Diffraction combined with tomography was applied for the monofilament reinforced systems as well. In this case in / ex situ tests showed thermal fatigue damage types and propagation under operation conditions.

All the samples used for the experiments in the following work are listed in Tab. 3.1. The first sample of AlSi7Mg/SiC/70p was commercially manufactured by Electrovac Austria [37] by gas pressure infiltration using their industrial production route of particle reinforced IGBT base plates [38]. The sample was cut from a baseplate in suitable dimensions (1.5 x 1.5 x 10 mm^3) for the simultaneous diffraction / tomography experiment at ID15A beam line at ESRF in 2006.

The particle reinforced Al-SiC and Al-diamond composites were produced by squeeze casting at EMPA Thun [4]. The coarse grained matrix of the slowly cooled squeeze cast Al matrix composites turned out to be inadequate for synchrotron diffraction (small gauge volume and low grain statistics). Therefore, neutron diffraction was applied and demanded a different sample size than the synchrotron tomography. Two samples of each of the composite types with dimensions, adequate to neutron and synchrotron experiments, were manufactured. The impossibility of mechanical post processing of diamond reinforced composites required sample infiltration in sizes suitable for the experiments. This challenging job was well done by our cooperating partners at EMPA Thun. Cylindrical samples in d ~ 6 mm, l ~ 10 mm for neutron diffraction and smaller ones in d ~ 0.8 mm and l ~ 10 mm for synchrotron tomography could be cast with almost perfectly homogeneous infiltration quality. The particle distribution was checked by conventional tomography at FH-Oberösterreich [39]. These samples were chosen for the scheduled experiment at the neutron and synchrotron sources. The particle size in the tomography samples was smaller, allowing a more homogeneous particle distribution over their 0.8 mm diameter, compared to the 6 mm neutron samples of larger particles with higher influence on stress evolution. One Al/CD/60p$_{ht}$ sample was heat treated at 640 °C for 5 h under pressure (remaining from squeeze casting) in order to study the effect of carbide formation at the interfaces on bonding strength.

The SiC fiber and W wire reinforced composites with pure copper matrix, Cu/SiC/20m and Cu/W/20m were produced by IPP Garching [40]. Firstly, the interfaces were deposited (magnetron sputtering) onto the raw fibers. Secondly, the Cu matrix was galvanically grown onto the fibers adequate in thickness to the desired volume fraction, and finally the fibers were sealed into a cylindrical capsule and consolidated by hot isostatic pressing (HIPing) at 650 °C with 100 MPa. All Cu/SiC/20m composites were delivered with SCS6 SiC fibers without coating and with a 200 nm Ti interlayer [31] for comparison. The MFRM samples with TiTaC coating were delivered with SCS0 SiC fibers [32] without protective graphite interlayer. The CuCr1Zr matrix composites were produced by DLR Köln [41]. A similar production route with matrix coated fibers was performed via HIPing, but at higher temperatures for the CuCr1Zr matrix at 800 °C and 150 MPa for several hours. The CuCr1Zr matrix containing 0.65 – 0.8 wt.% Cr and 0.07 – 0.15 wt.% Zr was soft condition after HIPing [42]. Precipitation hardening of the matrix was realized by additional solution treatment at 950 °C for 10 min, by quenching and then by age hardening at 470 °C for 60 min. The SiC filament reinforced CuCr1Zr composites were all delivered with SCS6 fibers. One uncoated fiber system with lowest volume fraction of 10 vol.% was tested comparable to 15 and 30 vol.% SiC both with a 200 nm Ti interlayer. The W wire reinforced Cu matrix composites were produced by IPP similar to the SiC fiber reinforced systems. W wires without interface treatment were investigated comparable to graded and etched interface designs. The W-Cu grading at the wire-matrix interface was realized by changing W/Cu concentrations during sputtering. Etched fibers were used in order to improve bonding by roughened wire-matrix interface. All fiber concepts were first sputtered by matrix, then galvanized and finally consolidated (HIPed) at the same temperatures as used for the SiC fiber reinforced systems. Two W reinforced MFRMs were tested, one of them with 30 vol.%, the other with 50 vol.%. The W wires were embedded by DLR directly into the matrix without interface treatment, assuming the bonding strength between W and CuCr1Zr to be sufficient. Conventional cylindrical tensile test samples were needed for neutron diffraction. The center part was cut in dimensions of $l \sim 16$ mm and $d \sim 3.5$ mm. Diffraction was made in two directions on the cylinder symmetric, unidirectional fiber reinforced MFRM samples.

Synchrotron tomography and diffraction were done on pure Cu matrix samples which were accordingly manufactured by IPP with a small fiber reinforced center part ($d \sim 1.2$ mm) and the surrounding matrix material turned off. The resulting MFRM samples had a dimension of $d \sim 1.6$ mm and $l \sim 12$ mm.

Composite	Matrix	Reinforcement	Type	Production	Dimension	Manufacturer
AlSi7Mg/SiC/70p	AlSi7Mg	SiC, 70 vol.% Ø ~ 5 – 200 µm	particle, trimodal	gas pressure infiltration	1.5 x 1.5 x 10 mm³ tomo, diff	Electrovac
Al/SiC/60p	Al	SiC, 60 vol.% Ø ~ 50 µm	particle, monomodal	squeeze casting	d ~ 0.8, l ~ 10 mm tomo d ~ 6, l ~ 10 mm diff	EMPA
Al/SiC/70p	Al	SiC, 70 vol.% Ø ~ 5, 50 µm	particle, bimodal	squeeze casting	d ~ 0.8, l ~ 10 mm tomo d ~ 6, l ~ 10 mm diff	EMPA
AlSi7/SiC/60p	AlSi7	SiC, 60 vol.% Ø ~ 50 µm	particle, monomodal	squeeze casting	d ~ 0.8, l ~ 10 mm tomo d ~ 6, l ~ 10 mm diff	EMPA
AlSi7/SiC/70p	AlSi7	SiC, 70 vol.% Ø ~ 5, 50 µm	particle, bimodal	squeeze casting	d ~ 0.8, l ~ 10 mm tomo d ~ 6, l ~ 10 mm diff	EMPA
Al/CD/60p	Al	Diamond, 60 vol.% Ø ~ 25 - 200 µm	particle, monomodal	squeeze casting	d ~ 0.8, l ~ 10 mm tomo d ~ 6, l ~ 10 mm diff	EMPA
Al/CD/60p$_{ht}$	Al	Diamond, 60 vol.% Ø ~ 25 - 200 µm heat treated	particle, monomodal	squeeze casting	d ~ 0.8, l ~ 10 mm tomo d ~ 6, l ~ 10 mm diff	EMPA
Al/CD$_{SiC}$/60p	Al	Diamond, 60 vol.% Ø ~ 25 - 200 µm SiC coated	particle, monomodal	squeeze casting	d ~ 0.8, l ~ 10 mm tomo d ~ 6, l ~ 10 mm diff	EMPA
AlSi7/CD/60p	AlSi7	Diamond, 60 vol.% Ø ~ 25 – 200 µm	particle, monomodal	squeeze casting	d ~ 0.8, l ~ 10 mm tomo d ~ 6, l ~ 10 mm diff	EMPA
AlSi7/CD$_{SiC}$/60p	AlSi7	Diamond, 60 vol.% Ø ~ 25 – 200 µm SiC coated	particle, monomodal	squeeze casting	d ~ 0.8, l ~ 10 mm tomo d ~ 6, l ~ 10 mm diff	EMPA
Cu/SiC/20m	Cu	SiC, 20 vol.% Ø ~ 140 µm	fiber, unidirectional	HIPing	d ~ 1.2, l ~ 16 mm tomo d ~ 3.5, l ~ 16 mm diff	IPP
Cu/SiC$_{Ti}$/20m	Cu	SiC, 20 vol.% Ø ~ 140 µm Ti coated	fiber, unidirectional	HIPing	d ~ 1.2, l ~ 16 mm tomo d ~ 3.5, l ~ 16 mm diff	IPP
Cu/SiC$_{TiTaC}$/20m	Cu	SiC, 20 vol.% Ø ~ 140 µm TiTaC coated	fiber, unidirectional	HIPing	d ~ 1.2, l ~ 16 mm tomo d ~ 3.5, l ~ 16 mm diff	IPP
CuCrZr/SiC/10m	CuCr1Zr	SiC, 10 vol.% Ø ~ 140 µm	fiber, unidirectional	HIPing	d ~ 3, l ~ 16 mm diff	DLR
CuCrZr/SiC$_{Ti}$/15m	CuCr1Zr	SiC, 15 vol.% Ø ~ 140 µm Ti coated	fiber, unidirectional	HIPing	d ~ 3, l ~ 16 mm diff	DLR
CuCrZr/SiC$_{Ti}$/30m	CuCr1Zr	SiC, 30 vol.% Ø ~ 140 µm Ti coated	fiber, unidirectional	HIPing	d ~ 3, l ~ 16 mm diff	DLR
Cu/W/20m	Cu	W, 20 vol.% Ø ~ 140 µm	wire, unidirectional	HIPing	d ~ 2, l ~ 16 mm tomo d ~ 3.5, l ~ 16 mm diff	IPP
Cu/W$_{etch}$/20m	Cu	W, 20 vol.% Ø ~ 140 µm Etched	wire, unidirectional	HIPing	d ~ 3.5, l ~ 16 mm diff	IPP
Cu/W$_{grad}$/20m	Cu	W, 20 vol.% Ø ~ 140 µm graded interface	wire, unidirectional	HIPing	d ~ 2, l ~ 16 mm tomo d ~ 3.5, l ~ 16 mm diff	IPP
CuCrZr/W/30m	CuCr1Zr	W, 30 vol.% Ø ~ 140 µm	wire, unidirectional	HIPing	d ~ 3, l ~ 16 mm diff	DLR
CuCrZr/W/50m	CuCr1Zr	W, 50 vol.% Ø ~ 140 µm	wire, unidirectional	HIPing	d ~ 3, l ~ 16 mm diff	DLR

Tab. 3.1.: List of the composite samples as used for diffraction and tomography experiments.

References

[1] Reddy GP, Gubta N, Material selection for microelectronic heat sinks: An application of the Ashby approach, Mat. and Des. vol. 31, 113-117, 2010.
[2] Huber T, Thermal expansion of Aluminium Alloys and Composites, PhD-thesis, Institute of Materials Science and Technology, TU-Vienna, 2003.
[3] Huber T, Degischer HP, Lefranc G, Schmitt T, Thermal expansion studies on aluminium-matrix composites with different reinforcement architecture of SiC particles, Comp. Sci. Tech. vol. 66, 2206-2217, 2006.
[4] http://www.epfl.ch
[5] Beffort O, Long S, Cayron C, Kuebler J, Buffat PA, Alloying effects on microstructure and mechanical properties of high volume fraction SiC-particle reinforced Al-MMCs made by squeeze casting infiltration, Comp. Sci. Tech. vol. 67, 737-745, 2007.
[6] Beffort O, Vaucher S, Khalid FA, On the Thermal and chemical stability of diamond during processing of Al/diamond composites by liquid metal infiltration, Diam. Relat. Mat. vol. 13, 1834-1843, 2004.
[7] Kleiner S, Khalid FA, Ruch PW, Meier S, Beffort O, Effect of diamond crystallographic orientation on dissolution and carbide formation in contact with liquid aluminum, Scri. Mat. vol. 55, 291-294, 2006.
[8] http://www.electronics-cooling.com
[9] http://www.epfl.ch
[10] http://www.ifam.fraunhofer.de
[11] http://www.plansee.at
[12] Schubert T, Ciupinski L, Zielinski W, Michalski A, Weißgärber T, Kieback B, Interfacial characterization of Cu/diamond composites prepared by powder metallurgy for heat sink applications, Scri. Mat. vol. 58, 263-266, 2008.
[13] http://www.extremat.org
[14] Mortensen A, Kelly A, Zweben C, Comprehensive Composite Materials, vol. 3, Metal Matrix Composites, 541-547, Oxford 2000.
[15] Battabyal M, Beffort O, Kleiner S, Vaucher S, Rohr L, Heat transport across the metal-daimond interface, Diam. Relat. Mat. vol. 17, 1438-1442, 2008.
[16] http://www.substech.com
[17] Zetterling CM, Process technology for silicon carbide devices, INSPEC, London, 2002.
[18] Knippenberg WF, Philips Research Reports, vol. 18, no. 3, 161-274, 1963.
[19] Yin LW, Li MS, Sun DS, Li FZ, Hao ZY, Some aspects of diamond crystal growth at high temperature and high pressure by TEM and SEM, Mat. Let. vol. 55, 397-402, 2002.
[20] Edtmaier C, Weber L, Tavangar R, Surface Modification of Diamonds in Diamond/Al-matrix composite, Adv. Mater. Res. vol. 59, 125-130, 2009.
[21] You, HJ, Bolt H, Overall mechanical properties of fiber-reinforced metal matrix composites for fusion applications, Jour. Nuc. Mat. vol. 305, 14-20, 2002.
[22] Herrmann A, Interface Optimization of Tungsten Fiber-Reinforced Copper for Heat Sink Applications, PhD-thesis, IPP, TU-München, 2009.
[23] Bevcar E, Aspekte der Kernfusionsforschung, Verlag der Österreichischen Akademie der Wissenschaften, Informationstagung April 1986.
[24] Stinchcomb WW, Ashbaugh NE, Composites Materials: Fatigue and Fracture, vol. 4, ASTM, 1916 Race Street, Philadelphia, PA 19103, 1993.
[25] Brendel A, Popescu C, Köck T, Bolt H, Promising composite heat sink material for divertor of future fusion reactors, Jour. Nuc. Mat. 367-370, 2007.
[26] http://www.specmaterials.com
[27] Peters P, Hemptenmacher J, Schurmann H, Production of MMCs on the basis of fibers coated with the metal alloy by magnetron sputtering, Mat. Sci. For. 931-935, 2007.
[28] Paffenholz V, Synthese und Charakterisierung von SiC_f/Cu-Matrix-Verbundwerkstoffe und ihre Anwendung in einem Modell einer Divertor-Komponente, PhD-thesis, IPP, TU-München, 2010.
[29] Brendel A, Popescu C, Leyens C, Woltersdorf J, Bolt H, SiC-fiber reinforced copper as heat sink material for fusion applications, Jour. Nuc. Mat. vol. 329-333, 804-808, 2004.
[30] Leyens C, Kocian F, Hausmann J, Kaysser WA, Materials and design concepts for high performance compressor components, Aero. Sci. Tech. vol. 7, 201-210, 2003.
[31] Brendel A, Woltersdorf J, Pippel E, Bolt H, Titanium as coupling agent in SiC fiber reinforced copper matrix composites, Mat. Chem. Phy. vol. 91, 116-123, 2005.
[32] Köck T, Brendel A, Bolt H, Interface reactions between silicon carbide and interlayers in silicon carbide-copper metal-matrix composites, Jour. Nuc. Mat. 362, 2007.
[33] Herrmann A, Schmid K, Balden M, Bolt H, Interfacial optimization of tungsten fiber-reinforced copper for high-temperature heat sink material or fusion application, Jour. Nuc. Mat. 386-388, 2009.
[34] Bolt H, Buuron A, Hemel V, Koch F, Gradient metal-a-C:H coatings deposited from dense plasma by a combined PVD/CVD process, Surf. Coa. Tech. vol. 98, 1518-1523, 1998.

[35] Ripoll MR, Ocenasek J, Microstructure and texture evolution during the drawing of tungsten wires, Eng. Frac. Mech. vol. 76, 1485-1499, 2009.
[36] Schöbel M, Requena G, Degischer HP, Kaminski H, Buslaps T, DiMichiel M, Residual stresses and void kinetics in AlSiC MMCs during thermal cycling, Mat. Sci. For. vol. 571-572, 413-418, 2008.
[37] http://www.electrovac.com/
[38] Lefranc G, Degischer HP, Sommer KH, Mitic G. Al–SiC improves reliability of IGBT power modules, In: Massard T, editor. ICCM12 proceedings, Paris, electronic support, 1999.
[39] http://www.fh-ooe.at
[40] http://www.ipp.mpg.de
[41] http://www.dlr.de
[42] Peters P, Hemptenmacher J, Schurmann H, The fiber/matrix interface and its influence on mechanical and physical properties of Cu-MMC, Comp. Sci. Tech. vol. 70, 9, 1321-1329, 2010.

4. Aims of investigations

Metal matrix composites which contain reinforcements with lower CTE (and higher E) than the matrix cause high internal micro stresses during changing temperatures that may lead to thermal fatigue damage at the interface and / or in the matrix. The understanding of the types of stresses, their magnitudes and effect on damage mechanisms under operation conditions would extend the knowledge of the requirements for material design.

a) Diffraction was used as non-destructive technique in order to investigate phase sensitive micro stresses under simulated operation conditions. Synchrotron as well as neutron radiation allow short time acquisition in deep penetrating diffraction measurements during thermal cycling. The lattice strains are determined relative to stress free references enabling stress calculation of the 3-axial stress state in a heterogeneous material. If stresses are present, the loading on the composites' constituents (matrix, reinforcement) can be observed. As soon as thermal fatigue damage occurs, it can be identified by a reduction of the stress amplitude during further repeated thermal cycles. The stresses as driving force for interfacial as well as thermal fatigue damage in the constituents can be measured in the different phases in the composite.

b) Synchrotron tomography is applied for non-destructive imaging of 3D reinforced systems i.e. PRM, IPC and MFRM in situ during thermal cycling. Thermal fatigue induced pores and cracks in the matrix, in the reinforcement or at the interfaces will be identified and distinguished. Defects from production such as infiltration and shrinkage voids in the matrix as well as cracks will be detected. Their effect on the measured stress levels and further thermal fatigue damage will be concluded.

5. Experimental background

5.1. Diffraction

For every diffraction experiment it is important to know how the diffraction pattern is formed and influenced by the unit cell and its periodic arrangement in space. The scattering theory bases on a Fourier transformation from real into reciprocal space. Structure, size and shape effects from real space on the diffraction pattern can only be understood with the following theoretical assumptions.

The principle of diffraction is elastic scattering of a wave on a solid body [1]. An incoming plane wave is scattered to a secondary wave (point source) in equ. 5-1.

$$\Psi(\vec{r}) = \Psi_0 \cdot e^{i\vec{k}\vec{r}} \cdot e^{-i\omega t} \tag{5-1}$$

with incoming wave Ψ_0, wave vector $\vec{k}$, space vector $\vec{r}$, angular velocity ω and time t.

The scattered wave is a sum over all outgoing spherical waves. The volume scattering strength is given by integration over the microscopic scattering centers (equ. 5-2).

$$\Psi(\vec{r}) = \frac{e^{ikR}}{R} \cdot \int_V \Psi_0(\vec{r}) \cdot f(\vec{r}) \cdot \frac{R}{\rho} \cdot e^{ik(\rho-R)} d^3r \tag{5-2}$$

with distance to the scatterer R, and scattering density in the scatterer ρ, and scattering factor $f(\vec{r})$.

Some approximations can be implemented for big distances R (equ. 5-3):

$$(\rho - R) \cong -\frac{\vec{R}}{R} \cdot \vec{r} \quad \text{and} \quad \frac{R}{\rho} \cong 1 \tag{5-3}$$

Equ. (2) can be thus simplified to equ. 5-4:

$$\Psi'(\vec{R}) = \frac{e^{ikR}}{R} \cdot \Psi_0 \cdot \underbrace{\int_V f(\vec{r}) \cdot e^{i(\vec{k}-\vec{k}')\vec{r}} d^3r}_{S(\vec{k},\vec{k}')} \tag{5-4}$$

The integral gives the scattering factor S. The measured quantity of the scattering cross section σ is the intensity of the scattered wave in a certain solid angle $d\Omega$ resulting in equ. 5-5.

$$d\sigma = \left|S(\vec{k},\vec{k}')\right|^2 d\Omega \tag{5-5}$$

This derivation holds for a single scatterer. In case of scattering from a solid a sum over N scattering centers has to be taken into account. The crystal is a periodical arrangement of microscopic scattering centers at the positions $r_{(j)}$ resulting in the scattering factor S for N scattering centers to equ. 5-6.

$$S(\vec{k},\vec{k}') = \sum_{j=1}^{N} f_j \cdot e^{-i(\vec{k}'-\vec{k})\vec{r}_{(j)}} \tag{5-6}$$

with j scattering centers of a scattering factor f_j.

Inelastic as well as multiple scattering events (dynamic scattering) are neglected in this first approach. The vector $\vec{k}'-\vec{k}$ is given by the reciprocal lattice vector $\vec{G}$ and an excitation error $\vec{s}$ (equ. 5-7).

$$\vec{G} + \vec{s} = \vec{k}' - \vec{k} \tag{5-7}$$

The sum over all scattering centers (atoms) in a crystal can be divided into two independent contributions (equ. 5-8).

$$S(\vec{k},\vec{k}') = \underbrace{\sum_{l=1}^{M} e^{-i(\vec{G}+\vec{s})\vec{R}_{(l)}}}_{lattice\ faktor\ (GF)} \cdot \underbrace{\sum_{j=1}^{n} f_j \cdot e^{-i(\vec{G}+\vec{s})\vec{r}_{(j)}}}_{strukture\ fakotr\ (F)} \tag{5-8}$$

The scattering factor is given now by a lattice factor GF and a structure factor F. The crystal includes M unit cells with n atoms per cell. The lattice factor GF only depends on the periodicity of the crystal where the structure factor F is influenced by the unit cell only. Some examples of shape influenced lattice factors GF are illustrated in Fig. 5.1.

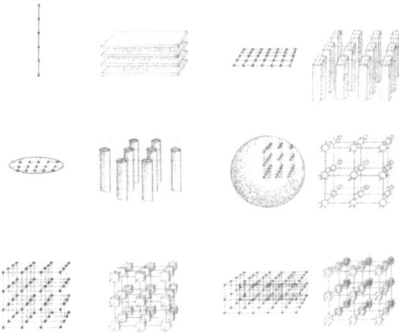

Fig. 5.1.: Examples of Fourier transformed shapes from real space structures in reciprocal space [2].

A diffractive measurement is done in real space, so that the Fourier transformation (equ. 5-9) of the space function can be observed. Real space dimensions and periodicities are reproduced in reciprocal space.

$$F(k) = \int_{-\infty}^{\infty} f(r) e^{ikr} dr \qquad (5-9)$$

with transformed real space function f(r), 1D real space coordinate r and reciprocal space coordinate k.

The macroscopic real space properties of the scatterer are projected into the spot shape in reciprocal space. The microscopic real space properties of the unit cell in real space (as discussed later) are reproduced by the periodicity of the diffraction spots in reciprocal space. The lattice factor GF is given by equ. 5-10

$$GF = \frac{1}{V_c} \cdot \int_V e^{-i\vec{s}\vec{r}} d^3r \qquad (5-10)$$

with unit cell volume V_c.

resulting in a δ-function for an infinite crystal. The maximum diffracted intensity appearing at s = 0 only represents the well-known Bragg relation in equ. 5-7 to equ. 5-11.

$$(\vec{k} - \vec{k}') = \vec{G} \qquad \text{and} \qquad \vec{k} \cdot \vec{G} = \frac{1}{2}|\vec{G}|^2 \qquad (5-11)$$

with reciprocal lattice vector $\vec{G}$.

This relation is valid if the wave vector $\vec{k}$ lies in the plane of symmetry of the reciprocal lattice vector $\vec{G}$. The plane of symmetry is also the border of the first Brillouin zone in the unit cell, which allows the construction of the Ewald sphere in the reciprocal space Fig. 5.2.

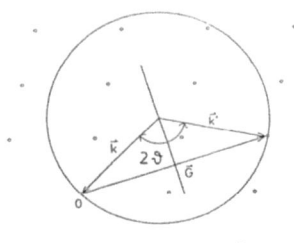

Fig .5.2.: The Ewald sphere for incoming wave $\vec{k}$ and outgoing wave $\vec{k}'$ the reciprocal lattice vector $\vec{G}$ and the diffracted angle ϑ[1].

The Ewald sphere passes through the origin of the reciprocal space O. In case that another reciprocal lattice point is lying on its surface the Bragg condition is fulfilled. The center of the sphere lays at $-\vec{k}$ and the radius is indirectly proportional to the wavelength of the diffracted wave function ($k = 2\pi/\lambda$). The Ewald sphere describes the Bragg equation in reciprocal space commonly known as:

$$n \cdot \lambda = 2 \cdot d_{hkl} \cdot \sin \theta \qquad (5\text{-}12)$$

with wavelength λ, periodic order n, lattice distance d_{hkl} and diffracted angle θ.

The arrangement of the diffraction spots in reciprocal space is given by the structure factor F from equ. 5-8 only. This structure factor depends on the atomic positions in the unit cell. Several types of crystal structures can be distinguished as illustrated in Fig. 5.3.

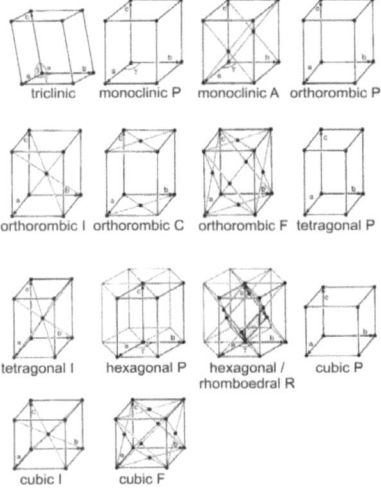

Fig. 5.3.: Some examples of the unit cells from the most common crystal structures with: primitive P, base centered C, face centered F, body centered I and side centered A [3].

The crystal structure is described by the size of its faces, their angles and the unit cell's type [1, 3]. The unit cells are described by the real space vectors $\vec{a_1}$, $\vec{a_2}$ and $\vec{a_3}$ in Fig. 5.4.

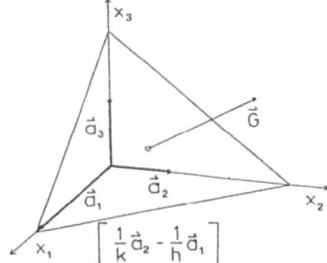

Fig. 5.4.: The reciprocal space vector $\vec{G}$ orthogonal to the real space elementary cell of the Bravais lattice [1].

A reciprocal space is introduced to project diffraction conditions in real space to a space of frequencies in reciprocal or k-space with reciprocal base vectors $\vec{A_1}$, $\vec{A_2}$ and $\vec{A_3}$ in equ. 5-13.

$$\vec{A_i} = 2\pi \frac{[\vec{a_j} \times \vec{a_k}]}{(\vec{a_i} \vec{a_j} \vec{a_k})} \tag{5-13}$$

with real base vectors $\vec{a_i}$.

This reciprocal unit cell volume is given by equ. 5-14.

$$V_c^* = \frac{(2\pi)^3}{V_c} \tag{5-14}$$

The reciprocal lattice vector $\vec{G}$ lies orthogonal to the corresponding lattice planes in real space as shown in Fig. 5.4. The length of $\vec{G}$ is given by the lattice distance d (equ. 5-15):

$$|\vec{G}| = \frac{2\pi}{d_{hkl}} \tag{5-15}$$

The Miller-indices {h, k, l} are introduced (equ. 5-16) in order to be able to describe the lattice distances and orientations in reciprocal space.

$$\vec{G} = h \cdot \vec{A_1} + k \cdot \vec{A_2} + l \cdot \vec{A_3} \tag{5-16}$$

Setting equ. 15 into equ. 5-8 results into the {h, k, l} dependent structure factor given by equ. 5-17.

$$F_{hkl} = \sum_{j=1}^{n} f_j \cdot e^{-2\pi i(u_j \cdot h + v_j \cdot k + w_j \cdot l)} \tag{5-17}$$

with reciprocal space vectors u, v, w.

This sum describes the intensities in k-space generated by the positions of the atoms in the unit cell. The characteristic diffraction patterns for each crystal structure (Fig. 5.3) are formed. These {h, k, l} with the structure factor F reduced to zero are called rules of extinction, which are significant for the different crystal types. If some atoms are periodically substituted like in a heterogeneous crystal (intermetallic), the changed atomic scattering factor may produce weak intensities at the positions of extinctions which are called superstructure.

5.2. Radiation types

Neutron and X-ray radiation is used for diffraction experiments depending on wavelength and interaction type. Not only the physical properties but also the possibilities of creation and manipulation of these probe particles / waves are limiting for their capabilities. Tab. 5.1 shows some fundamental properties of neutrons and X-rays which are desired as well as restricting for an experiment [3, 4].

Radiation	Particle	Spin	Mass	Scattering	Energy	Wavelength
X-ray	Photon	$s = 1$	$m_0 = 0$ kg	electromagnetic interaction	20 - 100 keV	~0.1 - 2 Å
Neutron	Nucleon	$s = 1/2$	$m_n = 1.7 \cdot 10^{-27}$ kg	strong interaction	25 meV (thermal)	~1 - 2 Å

Tab. 5.1.: X-rays and neutron radiation in comparison.

The neutron is a particle with a rest mass and can be interpreted as wave with its characteristic de-Broglie wavelength. The interaction with matter, necessary for any scattering event, is low due to its uncharged character. The neutron as fermion with a magnetic moment and strong interaction characteristics of a hadron only interacts with the nucleus and uncoupled electrons in the shells [5]. Therefore, the interaction with matter is low compared to X-rays. The X-ray photon without rest mass and magnetic moment interacts with the electrons in the shell, which represent a large number of scattering centers distributed around the lattice atoms.

5.2.1. X-rays

X-rays, discovered by Conrad Röntgen in 1895, are electromagnetic waves (photons) with an energy range from ~ 100 eV to the lower MeV spectrum [6]. The corresponding wavelengths (50 nm – 1 pm) cover the same dimensions as atomic distances in most of the known materials. X-rays were originally used in medical applications (absorption contrast imaging or radiography) and emerged as powerful tool for diffraction experiments in solid state physics and materials science over the last decade. Their interaction with electrons in an atom results in many possible scattering phenomena, which are important to describe mechanical and chemical properties of a material. Beside some inelastic scattering like photoelectric effect, fluorescence, Auger effect and Compton scattering, the elastic scattering is the basis for diffraction experiments as discussed here.

An X-ray wave interacts with electrons in an atom by inducing an oscillation which produces a secondary emitted wave, the diffracted wave (Fig. 5.5).

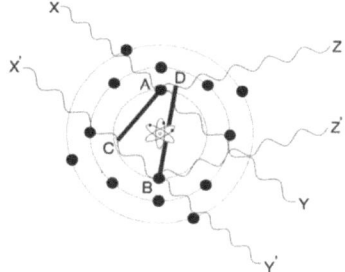

Fig 5.5.: X-ray scattering event with the electrons in an atom [7].

The atomic scattering factor from equ. 5-17 is a combination of several electronic scattering factors (equ. 5-18) [8].

$$f_e = \int_V e^{2\pi i(k-k')r} \rho(r) dV = \int_0^\infty 4\pi r^2 \cdot \rho(r) \cdot \frac{\sin(wr)}{wr} dr \qquad (5\text{-}18a)$$

$$w = \frac{4\pi \cdot \sin\theta}{\lambda} \qquad (5\text{-}18b)$$

The electron density $\rho(r)$ represents the electron cloud surrounding the nucleus as diffuse scattering center. This density results from the electron wave function taken from quantum mechanical calculations for each electron in each element (equ. 5-19).

$$\rho(r) = |\Psi|^2 = \frac{e^{-\frac{2r}{a}}}{\pi \cdot a^3} \qquad (5\text{-}19)$$

The last term in equ. 5-18a includes the estimated diameter a of the electron shell in the atom. The atomic scattering factor f is given by the sum of each electron scattering factors f_e (equ. 5-20).

$$f = \sum_n f_e = \sum_n \int_0^\infty 4\pi r^2 \cdot \rho(r) \cdot \frac{\sin(wr)}{wr} dr \qquad (5\text{-}20)$$

f is an amplitude of unmodified scattering per atom expressed in electron units. Since f is a function of $(\sin\theta/\lambda)$, the scattered intensity will fall off with theta. This fundamental property of X-ray diffraction is a consequence of electron cloud as scattering center with finite spatial expansion in magnitude similar to the wavelength. The larger the angle, the bigger the phase shift which reduces the scattering power [9].

Another fundamental scattering property of X-rays is the so-called Z-dependence of the scattering cross section. Z is given by the integration of the electron density over space (equ. 5-21).

$$\sum_n \int_0^\infty 4\pi r^2 \cdot \rho_n(r) dr = Z \qquad (5\text{-}21)$$

with $\rho_n(r)$ as electron density function for each n electron.

Insertion of the atomic scattering factor of equ. 5-18 with equ. 5-20 into the scattering cross section equ. 5-5 will result the well-known Z^2 dependence to equ. 5-22 [8].

$$\frac{d\sigma}{d\Omega} = |S(\vec{k},\vec{k}')|^2 \propto \left|\sum_n f_{en}\right|^2 \propto Z^2 \qquad (5\text{-}22)$$

5.2.2. X-ray sources

X-rays are produced by electron acceleration processes or inner shell recombinations in atoms. X-rays were first produced by the so called "Bremsstrahlen" effect where fast electrons hit a solid material and transfer their kinetic energy into an emitted photon [6]. This concept was further developed to the now commonly used X-ray tubes (Fig. 5.6).

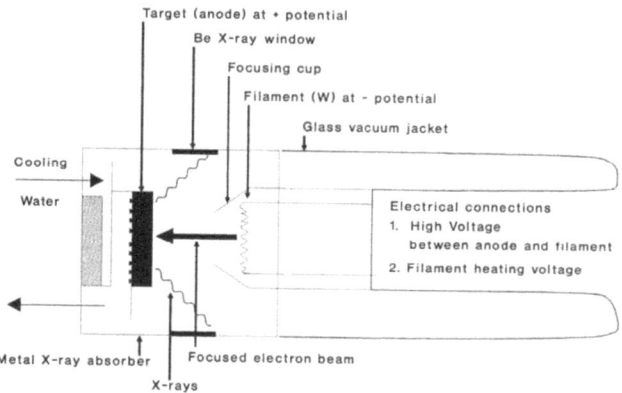

Fig. 5.6.: A X-ray tube and their components to accelerate electrons against a solid anode target [7].

In an X-ray tube, a filament is heated and emits thermally induced electrons into the surrounding vacuum. These free electrons get accelerated from the anode (filament) to the cathode by high voltage. X-rays, also known as "Kathodenstrahlung", are produced by electron scattering on the atoms (Bremsstrahlung) and by deflection of inner shell electrons of the anode material. An outer shell electron fills the free inner vacancy and emits an X-ray photon with a wavelength corresponding to the electron energy levels of the anode material. Its characteristic peak energy at the highest intense transition ($K\alpha_1$) is usually used to produce monochromatic X-rays required for laboratory angle dispersive diffraction experiments [7, 8]. The X-ray beam is transmitted through a low absorbing beryllium window out of the vacuum chamber. Additional absorption filters are externally applied which cut the $K\alpha_2$ of the cathode material to increase monochromaticity. Usually the next higher ordered element of the cathode material is used as filter (with $K\alpha_1$-filter at energies of $K\alpha_2$-cathode). The anode has to be actively cooled under operation, due to the high energy densities deposited in a small region of the anode. Rotating anodes are used in order to reduce the power density in the anode material by a volume increase. At first problematic for vacuum stability, due to the sealed rotation axis going from outside into the vacuum chamber, they are now operative in most X-ray tubes. Some new developments use transmission targets where samples can be placed very close to the X-ray source. These setups are emerging as micro spot sources for diffraction and imaging techniques.

Nowadays the most powerful X-ray sources are electron accelerators [10]. The fundamental idea of these facilities is to concentrate the X-ray emission into a small spatial angle using the Lorentz-transformed "Hertzscher Dipol" of a relativistic electron. The classical formulation of radiation of an accelerated particle is given by equ. 5-23 [11, 12].

$$P = \frac{2}{3}\frac{e^2}{m_0^2 c^3}\left(\frac{dp}{dt}\right)^2 \qquad (5\text{-}23)$$

with electron charge electron rest mass m_0 speed of light c and particle impulse p.

The angle distribution of radiated power (Hertzscher Dipol) as shown in Fig. 5.7 is given by equ.5-24.

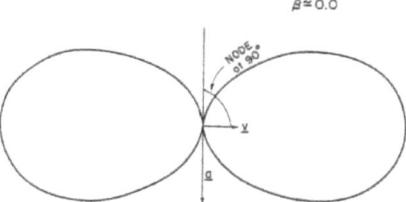

Fig. 5.7.: The irradiated intensity (Hertzscher Dipol) of a vertically oscillating charged particle [11].

$$\frac{dP}{d\Omega} = \frac{1}{4\pi}\frac{e^2}{\varepsilon_0 m_0^2 c^3}\left(\frac{dp}{dt}\right)^2 \sin^2(\theta) \qquad (5\text{-}24)$$

with irradiated power P, room angle Ω and azimuthal angle θ.

The same power distribution can be assumed for X-ray tubes with multiple sources (electrons) randomly orientated in space producing a diffuse light source. If a charged particle gets accelerated up to relativistic speeds, an energy dependence of the irradiated power distribution (spatial) has to be assumed [10]. The consequence of the Lorentz transformation of a linear accelerated particle is equ. 5-25.

$$\left(\frac{dp_\mu}{d\tau}\right)^2 \rightarrow \left(\frac{d\vec{p}}{d\tau}\right)^2 - \frac{1}{c^2}\left(\frac{dE}{d\tau}\right)^2 \qquad (5\text{-}25a)$$

with energy E, relativistic impulse p_μ, and dilated time τ.

$$\gamma = \frac{E}{m_0 c^2} = \frac{1}{\sqrt{1-\beta^2}} \quad \text{and} \quad \beta = \frac{v}{c} \quad (5\text{-}25\text{b})$$

with particle velocity v.

If the particle moves on a ring trajectory, following assumptions can be made (equ. 5-26):

$$\frac{dp}{dt} = p\omega = p\frac{v}{R} \approx p\frac{c}{R} = \frac{E}{R} \quad (5\text{-}26)$$

with trajectory's radius R.

The irradiated power of an accelerated particle in a storage ring is now given by equ. 5-27.

$$P = \frac{2}{3}\frac{e^2 c}{R^2}\beta^4 \frac{E}{m_0 c^2}^4 \quad \text{and} \quad \beta = \frac{v}{c} \xrightarrow{v \to c} 1 \quad (5\text{-}27)$$

This power represents the energy dissipation loss of a particle in a storage ring [10, 12]. It is clear that further particle energy increase of new storage rings requires to enlarge the radius of the facilities in order to keep the energy loss under operation in a feasible value. The particle rest mass is a critical value, for example an electron has a 10^{13} times higher irradiated power compared to a proton at same kinetic energies [10]. This is the reason why hadron accelerators are designed for particle physics, where high energy particles (with low radiation losses) are required, and why electron / positron accelerators are used as x-ray sources. In this case the higher irradiated power of an electron during storage is of minor importance compared to the possibilities of significantly higher X-ray power densities in the insertion devices (primary X-ray sources of a storage ring). The angular distribution of the power density of an accelerated particle in a synchrotron is given by equ. 5-28 [10].

$$\frac{dP}{d\Omega} = \frac{1}{4\pi}\frac{e^2 \dot{v}^2}{c^3}\frac{1}{(1-\beta\cos\psi)^3}\left[1 - \frac{\sin^2\psi \cos^2\phi_{xz}}{\gamma^2(1-\beta\cos\psi)^2}\right] \quad (5\text{-}28)$$

with space angles ψ and ϕ.

The two spheres of the dipol radiation from equ. 5-24 and Fig. 5.7 get Lorentz-transformed into particle moving direction tangential to its ring trajectory in the accelerator (Fig.5.8).

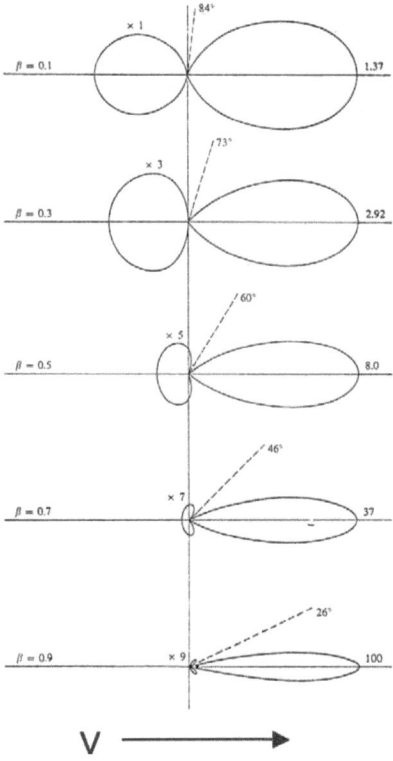

Fig. 5.8.: Lorentz-transformed dipol radiation of a relativistic particle [11].

The natural opening angle of the irradiated power depends on β and decreases with increasing particle energies. In first estimation the opening angle is correlated to the particle energy by (equ. 5-29).

$$\vartheta \cong \frac{1}{\gamma} \quad \text{and} \quad \gamma = \frac{E}{m_0 c^2} \quad (5\text{-}29)$$

with relativistic total dipol-opening angel ϑ.

Therefore the ring radius of further accelerators is increased to improve the X-ray beam collimation of a particle at higher energies.

The X-ray sources in storage rings are the so called insertion devices which accelerate the particles transversal to their trajectory [10, 13]. Combinations of di-, quadro-, hexa-, and octo-pol magnets are used to deflect, focus or defocus the relativistic particle current in the synchrotron. The simplest form of an insertion device (ID) is the so called bending magnet which represents a dipol magnet inducing a field transversal to the ring trajectory. The most powerful synchrotron X-ray sources are wigglers and undulator devices consisting of multipol magnet arrays arranged around a straight segment of the electron trajectory (Fig. 5.9).

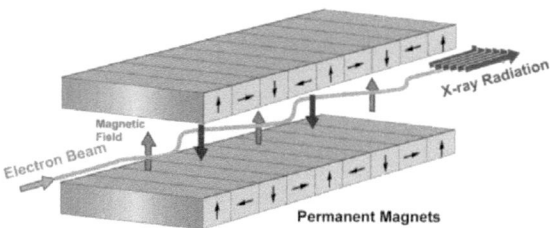

Fig. 5.9.: A multipol magnet induces an oscillation of the relativistic particles transversal to their propagation direction in the accelerator [14].

To distinguish between wiggler and undulator the K factor is introduced in equ. 5-30.

$$K = \frac{eB_0}{2\pi} \frac{\lambda_{ID}}{m_0 c} \qquad (5\text{-}30)$$

with magnetic field B_0, and insertion devices periodic length λ_{ID}.

For K < 1 the oscillating electron trajectory is smaller than the beam opening angle ϑ which is called an undulator. In a wiggler with K >> 1 the electron acts as separated source on each position on the trajectory (Fig. 5.10).

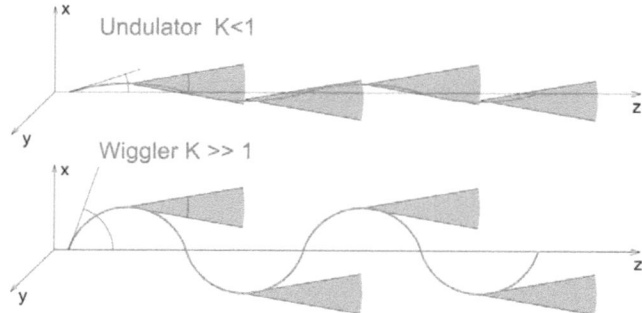

Fig. 5.10.: The electron trajectory relative to the natural X-ray beam opening angle [12].

The wiggler (Fig. 5.10) is a N dipol-source, with N as the number of magnets in the array, giving an intensity proportional to **N·P** (with P as dipol source). The radiation emitted by the wiggler forms a continuous X-ray spectrum (Fig. 5.11) below and above the critical wavelength of the storage ring (equ. 5-31).

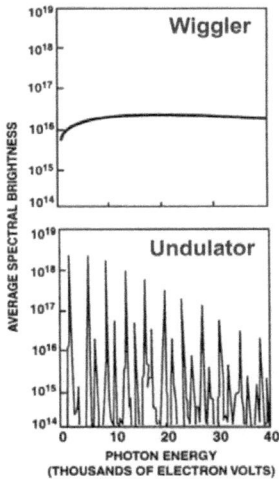

Fig. 5.11.: A typical wiggler and undulator X-ray wavelength distribution [11].

$$\lambda_c = \frac{4\pi}{3}\gamma^{-3}R \qquad (5\text{-}31)$$

with storage ring radius R.

The critical wavelength of a synchrotron gives the median value of spectral intensity. The synchrotron has a broad spectrum above this wavelength but an exponential decay below. The wavelength of the emitted photons is limited by the particle energy.

The undulator (Fig. 5.11) with its small K < 1 produces interference effects from the X-rays emitted at N dipoles within the beam opening angle ϑ. The resulting intensity distribution depends on the undulators period λ_u (equ. 5-32).

$$\lambda = \frac{\lambda_u}{2\gamma^2}\left(1 + \frac{K^2}{2} + \gamma^2\theta^2\right) \qquad (5\text{-}32)$$

with undulator's period length λ_u and opening angle θ.

Also the resulting total opening angle ϑ is reduced from the natural dipol radiation equ. 5-29 to equ. 5-33.

$$\vartheta \cong \frac{1}{\sqrt{N}\gamma} \qquad (5\text{-}33)$$

with a N-value of ~ 50-100. The result is a highly brilliant X-ray beam of an intensity relation equal to $N^2 \cdot P$, but only discrete energy values corresponding to λ_u.

For all insertion devices the X-ray intensity depends inherently on the current in the storage ring i.e. the number of stored electrons. The number of photons N is given by equ. 5-34.

$$N\left[\frac{photons}{s \cdot eV \cdot mA \cdot mrad}\right] = 4.5 \times 10^{12} \frac{j[mA] \cdot \sqrt[3]{R[m]}}{\varepsilon[eV]^{-\frac{2}{3}}} \qquad (5\text{-}34)$$

with number of N, particle beam current J, trajectory radius R and particle energy ε.

From this equation the critical photon energy and wavelength dependence on the energy of the storage ring can be derived. The particle current in modern synchrotrons is not continuous but pulsed. These electron ensembles travelling at relativistic velocities in the accelerator are called bunches and have typical pulse lengths of 50 ps to 1 ns. It is obvious that for further photon intensity, wavelengths and beam brilliance improvements storage ring design goes into higher energy and current regimes. The larger radius of such devices reduces the permanent radiation losses.

5.2.3. X-ray detectors

The simplest method of detecting X-rays is the photographic film [7]. When exposing silver halide (AgX) grains in a film to X-rays (same as for visible light) they get reduced to pure silver on the bottom of the gelatin substrate. The darkening of the photographic film goes linear with its absorbed intensity (if not saturated). The spatial resolution and sensitivity of these films is quite high and allows good acquisition of static diffraction patterns, used for line profile analysis or powder diffraction. Unfortunately, the film once exposed remains in this acquired condition so that the death time of a photographic film for X-ray detection is infinite.

Photoemissive and photoconductive detector systems base on the concept of interaction of a photon with gas or solid material [10]. The principle of photoemissive detectors is the release of a photoelectron in a medium which produces a current between two electrodes. An external voltage produces a current of photoelectrons which is proportional to the absorbed X-ray intensity. The energy of the X-rays has to be high enough to free an electron in the absorbing medium which is defined by the medium's work function.

Gas ionization chambers represent a photoconducting system where the incoming photon ionizes gas [7]. Usually the reaction chamber is filled with argon, where a Ar^+ and e^- couple is formed and produces a current between the electrodes (Fig. 5.12).

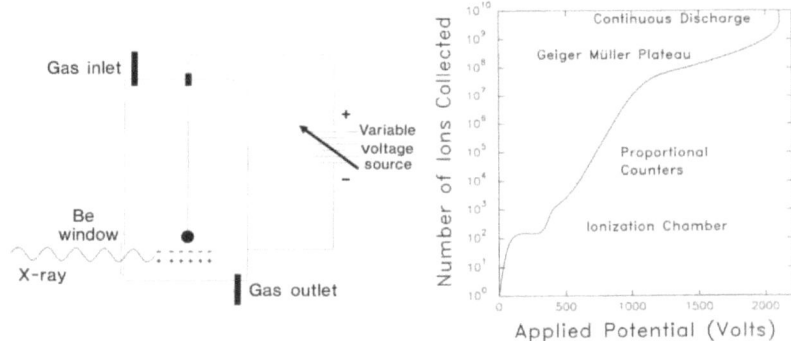

Fig. 5.12.: Gas ionization chambers operated in different voltage regimes [7].

A gas ionization chamber can be operated in three different modes, either as ionization chamber, or as proportional counter, or as Geiger-Müller counter. For the ionization counter, the externally applied potential is low, and the current depends on the X-ray intensity only. For each X-ray photon with E(photon) > E(ionization) one ion-electron pair is produced. The proportional counter with a higher applied voltage produces secondary electrons which interact with the next Ar-atoms by further ionization. The current depends now on the excessive energy of the photon > E(ionization) which goes into the kinetic energy of the secondary electrons. The higher this energy rises the more secondary ionization will happen. The current in a proportional counter is therefore proportional to the energy of the incoming X-ray. The third well-known setup of this detector principle is the so called Geiger-Müller detector operating at a high potential above ~ 1kV far from the linear region of the ionization chamber. Each absorbed X-ray photon will cause complete discharge of the chamber by an avalanche effect which further ionizes almost all Ar atoms. No energy or intensity can be resolved but the radiation to current relation is drastically increased. A good current signal is caused by every detected photon even if low powers are disposed into the absorbing medium. A consequence will be a high dead time (time needed to recombine the ions) of ~ 200 μs.

The most widely used example of a photoconducting detector is the scintillation counter (Fig. 5.13) [10]. The incoming X-ray is absorbed by an alkali-halide crystal (NaI doped with Tl atoms). The high-energy electrons produced by the X-ray photons excite several secondary electrons into the conduction band of the crystal. These electrons get captured by the trace impurities (Tl) and produce fluorescence photons (scintillas) in the visible light spectrum of
$\lambda \sim 0.4$ μm. If this "blue" light further reacts with the cesium-antimony photocathode, secondary electrons are produced and accelerated onto the next lower charged cathode. An amplification effect is caused by multiplication (photo-multipler) of the accelerated electrons at every cathode interaction in the photo-multiplying tube.

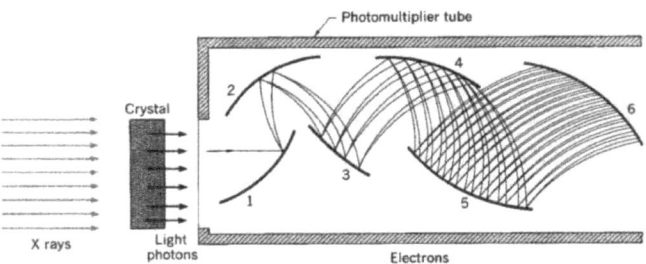

Fig. 5.13.: A scintillator counter with photo-multipler tube on the back [2].

The electrical signal of the collected electrons is proportional to the energy of the incoming wave. Unfortunately, the large number of losses reduces the energy resolution limit. The scintillator is used for laboratory X-ray diffraction as a 1D counter in a monochromatic angle dispersive diffraction setup (powder diffraction, spectral scanning and stress measurements). High quantum efficiency and low dead time ~ 1 μs are useful in this application.

Another technique of photoconducting detectors in the high energy photon regime of synchrotron radiation utilizes Si- or Ge-semiconductor crystals doped with Li [10]. In this case, the photons excite electron-electron hole pairs which produce a current in the semiconductor depending on energy and intensity of the incoming X-rays. The Si(Li) system provides a typical energy resolution of ~ 140 eV at 1 keV. These detectors have to be cooled to -170 °C in order to reduce thermal noise for background reduction. The Ge(Li) detectors provide better energy resolution (~ 2 eV) at higher energies (~ 50 keV) but have to be kept constantly at -170 °C to avoid intrinsic Li-compensation in the Ge crystal. The spatial resolution is not in the range of CCD systems as will be discussed later. Therefore they are usually applied as 1D-energy dispersive X-ray detector systems (EDS). The energy resolution limiting for diffraction is lower, compared to new 2D-detector systems, as well.

For acquisition of 2D diffraction images photoconducting and photoemissive PSD (Position Sensitive Device) detectors were developed. As photoconducting example, a 2D array of Si-diodes (similar the semiconducting EDS devices) absorbs X-ray photons that produce a current in defined regions. Intensity and location of the incoming photons project an image of the diffracted beam in a certain interval in space. These systems are cooled to reduce their leakage current which is important for image collection (counting). The resolution is defined by the pixel-size of the diodes in the array and depends on the distance to the scattering center (sample). Comparable PSD systems apply the photoemissive technique like micro channel array plates. These micro channels absorb photons on a photocathode which releases a photoelectron into a channel oriented (often slightly bended) into beam direction (Fig. 5.14). The electron gets accelerated in this channel exciting more secondary electrons when hitting the walls. The electrons are collected by an array on the bottom which detects the current and the position of the signal comparable to the semiconducting devices.

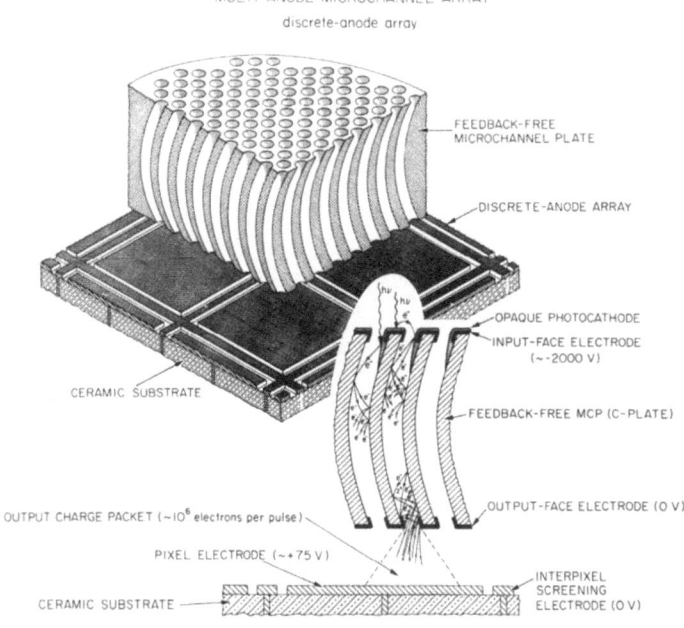

Fig. 5.14.: A micro channel PSD detector array for 2D signal acquisition [10].

The charge coupled devices (CCD) are the 2D photon detectors with the highest spatial resolution. The CCD consists of a 2D MOS capacitor array with overlapping depletion regions (Fig. 5.15).

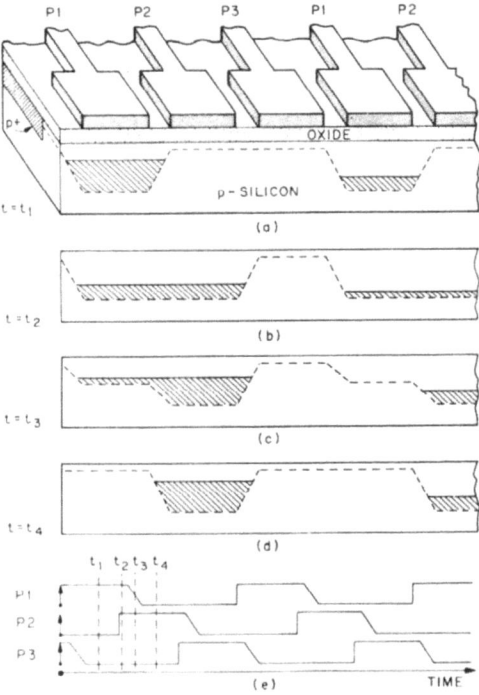

Fig. 5.15.: Sketch of 3-phase-channel CCD with charge carrying potential wells in the silicon substrate. The potential wells travel below the capacitor array at subsequent time intervals from (a) to (d).

The charge can be moved stepwise from element to element in one dimension of the array allowing a high lateral resolution. Unfortunately these systems are limited to the long wavelength region of visible light down to the UV region. Therefore an X-ray beam has to be transformed into visible light before detected at the 2D image array. These CCDs are used for X-ray imaging, as described later, with lateral resolutions down to the sub-μm regime.

A powerful method of collecting 2D images of X-ray diffraction patterns is the image plate detector. This system contains a round image plate coated with a Eu^{2+} doped BaBrF layer. If this layer is exposed to X-rays the oxidation process to Eu^{3+} stores intensity information local distributed on the layer. After acquisition the plate is rotated and helical read out by a laser. The image is erased by exposal to visible light. The image plate detector reaches resolutions in the μm regime and is comparably stable to temperature with good resistance to beam damage. Acquisitions in seconds are possible, but the readout time of ~ 100 sec is limiting for fast sequential acquisition.

5.2.4. Neutrons

A neutron beam is a particle radiation consisting of heavy uncharged nucleons with magnetic moment (spin = ½) and kinetic energy proportional to their temperature [4]. With the De-Broglie relation these particle radiation can be assumed as wave field in space with defined wavelength (equ. 5-35).

$$E = \frac{m_n v^2}{2} = k_B T = \frac{(\hbar k)^2}{2m_n} \tag{5-35a}$$

$$k = \frac{2\pi}{\lambda} = \frac{m_n v}{\hbar} \rightarrow \lambda = \frac{h}{p} \tag{5-35b}$$

The neutron temperature is a degree of the amount of kinetic energy (Tab. 5.2) [3]

Type	Energy (meV)	Temperature (K)	Wavelength (Å)
hot	0.1 - 10	1 - 120	4 - 30
thermal	5 - 100	60 - 1000	1 - 4
cold	100 - 500	1000 - 6000	0.4 - 1

Tab. 5.2.: The different neutron energy regimes.

Thermal neutrons with a wavelength similar to the interatomic spaces in solid materials are used for diffraction experiments. In contrast to X-rays with their electromagnetic interaction with the electron cloud of the atoms, the uncharged neutron interacts with the nucleus only (neglecting magnetic spin interactions with uncoupled electrons). These nuclear scattering centers are point like sources compared to the relatively large electron shells in X-ray diffraction. Energies of thermal neutrons (~ 25 meV) are too low to change energies in the nucleus, therefore only elastic scattering can be assumed (beside absorption resonances in some nuclei). The wave lengths $\lambda \sim 10^{-10}$ m are big comparable to the nucleus diameter $\emptyset \sim 10^{-15}$m, and ideal scattering centers without any shape effect can be assumed [3]. These two properties combined make thermal neutrons excellent probe particles for diffraction on crystalline structures [9].

The Born approximation for weak scattering events ($\sigma/a^2\pi \ll 1$) can be used to describe neutron scattering on a crystal due to the size of the scattering center a (nucleus) and the neutron scattering cross section σ resulting in $\sigma/a^2\pi \sim 10^{-7}$[3]. The first Born term with higher orders neglected (only single scattering event estimated) gives the scattered neutron amplitude to equ. 5-36.

$$f_N = -\frac{1}{4\pi}\int_V e^{-i(\vec{k}-\vec{k}')\vec{r}}U(r)d^3r \qquad (5\text{-}36)$$

The incoming wave $\vec{k}$ is diffracted on a potential described by the evolution operator U(r) to the scattered wave $\vec{k}'$. The scattering cross section can be calculated as in equ. 5-5, 5-6 by equ. 5-37 introducing the Fermi pseudo potential V(r).

$$d\sigma(\theta,\varphi) = |f_N(\theta,\varphi)|^2 d\Omega \qquad (5\text{-}37a)$$

$$U(r) = \frac{2\mu}{\hbar}V(r) \qquad (5\text{-}37b)$$

The differential scattering cross section can be given proportional to the potential (equ. 5-38):

$$\frac{d\sigma}{d\Omega} \propto \left|\int V(r)e^{i(\vec{k}-\vec{k}')\vec{r}}dr\right|^2 \quad \text{and} \quad V(r) = \frac{2\pi\hbar^2}{m_n}\cdot b\cdot\delta(r-R) \qquad (5\text{-}38)$$

Neutron diffraction on a crystal is described by scattering of a particle wave on point like scattering centers. The Fermi pseudo potential with its delta function is not zero only at the atomic positions, and includes the neutron scattering length b with a real and complex contribution (equ. 5-39).

$$b = b_1 + ib_2 \qquad (5\text{-}39)$$

Here b_1 gives the scattering and b_2 the absorption. In neutron diffraction this scattering length is not systematic and does not depend on Z (contrary to X-ray scattering). The value differs between the elements and their isotopes and can be negative in some cases (H, Ti, Mn, Ni). Negative neutron scattering lengths means a resonance in the nucleus, where the primary neutron is absorbed forming a compound nucleus emitting a secondary neutron with high incoherent contributions. In case of a positive scattering length, the scattering results from a full elastic potential interaction between neutron and nucleus (strong interaction) with a 180° phase shift in the scattered wave function. A negative scattering length produces no phase shift.

5.2.5. Neutron sources

Free neutrons are instable particles (halflife ~ 10 min), rarely appearing in terrestrial environments and difficult to produce. No acceleration processes can be applied compared to synchrotron radiation due to their uncharged character [3, 4]. All available methods are diffuse sources similar to X-ray tubes on the photon sector [15].

The common method of neutron production is fission of ^{235}U [3]. A thermal neutron (25 meV) hits a nucleus and gives rise to vibrations which result in a disintegration into two unequal nuclei (Fig. 5.16).

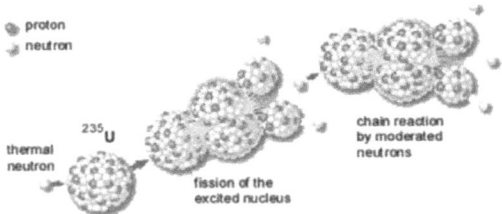

Fig. 5.16.: The fission process of ^{235}U for neutron production [15].

During fission of the uranium nucleus 2 to 3 fast neutrons (several MeV) are produced. The fast neutrons have to be moderated by D_2O down to the thermal energy regime before reaction with the next uranium nucleus can happen (chain reaction). These thermal neutrons are required for diffraction experiments as well. In average one neutron has to be left in order to keep the nuclear reaction operative. Every 2nd neutron is extracted for the beam. The moderation of the hot neutrons requires a slowing down length L_s of ~ 29 cm in the D_2O tank. If a point source of fast neutrons is assumed, the initial neutron flux decreases as a factor of 10^{-4} due to L_s. Further monochromatization, collimation and shielding reduce the final intensity on the sample by the factor of 10^{-11} to 10^{-17} depending on the $\Delta\lambda/\lambda$ resolution requirements. Therefore, good monochromatism is often sacrificed for higher intensity by using only one focusing single crystal monochromator. Beside the common continuous fission sources pulsed reactor concepts were developed as well, in order to increase the neutron flux during short time periods. These sources were not further developed due to the required instability of the nuclear reaction under pulsed conditions.

A more powerful method is the pulsed spallation source [3]. High energy protons (> 400 MeV) react with nuclei which release large amount of neutrons, protons, mesons and gamma radiation. The reaction called spallation is illustrated in Fig. 5.17.

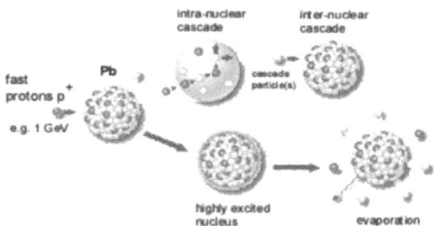

Fig. 5.17.: The spallation process of heavy nuclei for neutron production [15].

The protons get accelerated in a linac (linear accelerator) and hit heavy atoms in the target inducing an inter-nuclear cascade. For example, at the ESS (European Spallation Source) proton energies of 1.33 GeV and currents of 3.75 mA contribute to a beam power of 5 MW. The highly excited evaporating nuclei are an intense point source of high energy neutrons. Heavy elements are favored as target material like Ta, W, Re, Pb, Bi or even U. The target has to be cooled due to the high energies deposited during proton beam impact. Also rotating targets are used to increase the exposed volume and to reduce the power density (same principle as rotation anode X-ray tubes). In spallation sources the neutrons have to be moderated to the thermal energy regime as well. In this case a H_2O moderator is preferred due to its higher absorption and smaller dimensions needed. The thermal storage time (lower for H_2O) is of main interest for good neutron peak flux. Small (~ 1.5 liters) moderators with a slowing down length L_s of ~ 18 cm are used.

Some examples of the neutron flux in continuous as well as in pulsed reactors and of spallation sources are given in Tab. 5.3.

	ILL Grenoble (50 MW)	FMR2 Garching (20 MW)	ESS	SNS Oak Ridge
peak flux [n s^{-1} cm^{-2}]	$1.2 \cdot 10^{15}$	$8 \cdot 10^{14}$	$1.4 \cdot 10^{17}$	$1.7 \cdot 10^{16}$
integral flux [n s^{-1} cm^{-2}]	$1.2 \cdot 10^{15}$	$8 \cdot 10^{14}$	$0.6 \cdot 10^{15}$	$0.17 \cdot 10^{15}$
pulse repetition [ms]	-	-	50	50

Tab. 5.3.: The integral and pulsed intensity properties of the most powerful neutron sources [3].

5.2.6. Neutron detectors

Neutrons cannot be detected like other charged particles or photons by their electromagnetic interaction with electrons. The uncharged neutron reacts with the particles in a nucleus driven by the strong interaction. Examples of neutron reactions with interactive elements are shown in Tab. 5.4.

Reaction	Cross section (25 meV)	Particles generated	Energy [MeV]
n + ^{3}He	5333 b	p, ^{3}T	0.77
n + ^{6}Li	941 b	^{3}T, ^{4}He	4.79
n + ^{10}B	3838 b	^{4}He, ^{7}Li, γ	2.3
n + ^{235}U	681 b	fission	1 - 2

Tab 5.4.: Interactions of neutrons with some elements used for neutron detection.

The most often used neutron detector is the ^{3}He proportional counter [3, 16, 17]. The big scattering cross section of ^{3}He allows good neutron detection with a ^{3}He-gas counter. The neutron absorption products ionize the gas atoms in the chamber. An external applied voltage accelerates these charged particles towards the electrodes, inducing a current which can be measured. The concept of a ^{3}He proportional counter is similar to the Ar proportional counter for X-ray detection only different in the radiation absorbing effect. In angle dispersive neutron diffraction these ^{3}He detectors are arranged in arrays of small gas capillaries to a PSD. A small design of these tubes allows pixel sizes of ~ 1.4 x 1.4 mm² over a 350 x 350 mm array delivering good spatial resolution for segmental scanning of peak profiles in angle dispersive neutron diffraction.

Scintillation counters similar to X-ray detection are used for neutrons as well [3]. The absorband for neutrons is ^{6}Li-glass which produces photons from electron recombination. These photons get detected, and their signal is amplified in a photomultipler. Same concept as described for X-ray diffraction uses a Ce doped Li responsible for these scintilla-photons. Other configurations base on LiI or ZnS crystals.

Semiconducting detectors are made of Si crystals doped with receptor elements such as ^{6}Li [3]. These detectors act as solid state ionization chambers where the neutron absorption products generate electron-hole pairs inducing a detectable current in the semiconductor matrix. The main achievement of solid state detectors is feasibility of small sized detection units in order to increase the spatial sensitivity of PSD systems. The dead time required for the recombination of load carriers is much more reduced compared to gas ionization chambers. This property gets important for the new pulsed spallation sources where efficient detection of short high energy neutron pulses will be demanding for the detectors systems.

The last detector system remained to be mentioned are fission detectors. Their heavy nuclei such as ^{235}U or ^{239}Pu react with a thermal neutron in a fission process into two lighter nuclei. In general these detectors with low counting probability are used to control beam stability or for monitoring.

5.3. Radiation properties

X-ray as well as neutron radiation is used for diffraction applications in solid state physics and materials science [4, 9]. These two competing radiations show entirely different characteristics of interaction with matter and physical properties: The photon (spin = 1, boson), an electromagnetic radiation without rest mass always travelling at the speed of light with wavelength dependent on its energy; The neutron a particle radiation (spin = ½, fermion), uncharged with rest mass and a wavelength dependent on its velocity (De-Broglie). The electromagnetic interacting X-ray is scattered by electrons in the shells or conduction bands, periodically distributed over the crystal. The comparably weak interacting neutrons react with the nucleus only, due to their uncharged character. X-ray scattering potential Ø ~ 10^{-10} m compared to neutron scattering potential
Ø ~ 10^{-15} m. Different penetration depths will be the result for the two types of radiation in Fig. 5.18.

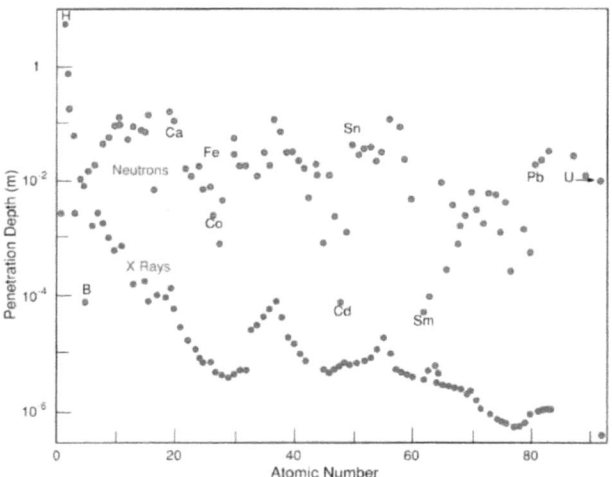

Fig. 5.18.: Penetration depth of neutrons compared to X-rays [4].

This relation shows the Z^2 dependence of the X-ray scattering strength and the non-systematic phenomenological neutron scattering cross section. The high reactivity of X-ray with photons results in high diffracted intensities but at the same time reduces the penetration depth significantly (especially for metals). The weaker interacting neutron penetrates 10^4 times deeper into the material beneficial for most engineering applications (in bulk measurements, gauge volume). The lower interaction property goes with lower signal intensity owing to longer acquisition of bigger gauge volumes required (lower spatial resolution possible). The entirely different neutron scattering cross sections of neighboring elements (not Z^2 dependence as for X-rays) makes neutrons and X-rays to complementary probe radiations in some aspects of diffraction as well as imaging.

Not only the radiation properties but also the availability of sources is a key issue for experimental demands. Some X-ray sources are compared with a typical high flux neutron source in Tab. 5.5.

	Brightness ($s^{-1}\,m^{-2}\,ster^{-1}$)	dE/E (%)	Divergence ($mrad^2$)	Flux ($s^{-1}\,m^{-2}$)
Neutrons	10^{15}	2	10 x 10	10^{11}
Rotating Anode	10^{16}	3	0.5 x 10	5×10^{10}
Bending Magnet	10^{24}	0.01	0.1 x 5	5×10^{17}
Wiggler	10^{26}	0.01	0.1 x 1	10^{19}
Undulator (APS)	10^{33}	0.01	0.01 x 0.1	10^{24}

Tab. 5.5.: Beam parameters of modern neutron sources compared to some X-ray sources [4].

As mentioned above, brightness and flux of modern neutron sources are comparable with a diffuse X-ray tube as photon source. The advantages of accelerators with relativistic beam collimation are not applicable for neutrons. As a consequence the modern X-ray sources on storage rings have not only a 10^{10} to 10^{18} brighter beam, but also 100 times lower energy blurring and beam divergence. The capability of modern photon sources lies far beyond modern neutron sources, due their highly collimated point source character compared to diffuse thermal "cloud" sources of neutrons.

Comparison:

X-ray radiation can be produced in highest brilliance by synchrotron excitation enabling measurements at short time, real and reciprocal space resolution.

The strong scattering photon reduces the penetration depth to mm dependent on Z^2 of the material (light elements are transparent, $Z < B$).

The scattering center of photons is the electron cloud which is relatively large (comparable to the interatomic spacing) with a diameter of $\sim 10^{-10}$ m resulting in a θ-dependence of the scattering power.

Neutron radiation is limited to diffuse thermal sources with 10^{13} lower flux compared to X-rays. Beam focussing and single monochromator crystals (required to keep flux losses as low as possible) increase the $\Delta\lambda/\lambda$ error.

The uncharged neutron interacts with the nucleus only (beside magnetic interactions) and penetration depths of several cm can be achieved. The scattering cross section is not Z dependent (bad for some elements: H, Li, Ti, Mn, Ni, Sm).

The neutrons interact with the nucleus and have a point like scattering center with a diameter of $\sim 10^{-15}$ m. Thermal neutrons are not influenced by geometric effects of the scatterer due to a 10^5 times bigger wavelength comparable to the nucleus' size.

5.4. Monochromatic diffraction on a polycrystalline material

Diffraction experiments on engineering materials deal with X-ray or neutron scattering on a complex polycrystalline often heterogeneous structure [8]. The discrete Laue-diffraction patterns of single crystallites get averaged to Debye Scherrer cones (known from powder diffraction). The intensities are a contribution of every crystallite randomly orientated in space fulfilling the instrumental Bragg condition. The Ewald sphere of a polycrystalline sample can be constructed as shown in Fig. 5.19.

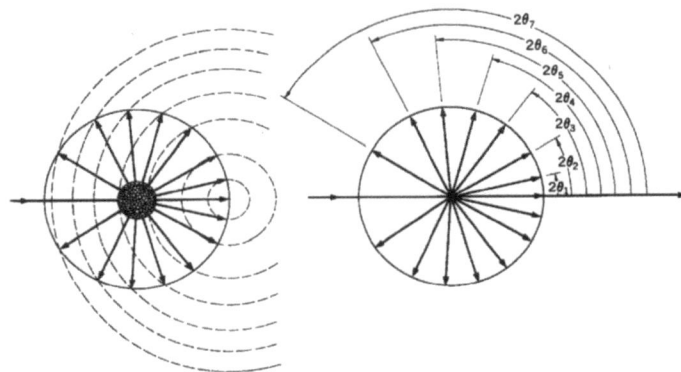

Fig. 5.19.: The Ewald sphere of polycrystalline materials corresponding to Fig. 5.2 [2].

The diffracted power of an {h, k, l}-reflection in such geometries can be calculated by integration over all orientations and the area of the projected surface (equ. 5-40) [8].

$$P = \iiint I_p \frac{Mm}{2} \cos\theta \, d\alpha R^2 d\beta d\gamma \qquad (5\text{-}40)$$

with space angles α, β, γ, diffracted angle θ and distance to the scatterrer R.

The intensity I_p of a single crystal is calculated from a single scatterrer diffracted intensity I and the structure factor F with the spatial dilatation N_1, N_2 and N_3 along the crystal axes a_1, a_2 and a_3 respectively (equ. 5-41).

$$\left(I_p\right)_{max} = I_e F_{hkl}^2 N_1^2 N_2^2 N_3^2 \qquad (5\text{-}41)$$

The number of crystals M (equ. 5-40) in the sample (gauge volume) contributes directly to the quality of diffraction pattern (Debye Scherrer ring continuity). As a consequence the choice of gauge volume size is not only a matter of spatial resolution versus diffracted intensity but also of the grain size of the material. Small gauge volumes are applicable for small grain sized materials whereas big gauge volumes are required for large grain sizes to ensure good grain statistics essential for "polycrystalline" diffraction. The second term implements the so called multiplicity m_{hkl} of a lattice in the unit cell. These multiplicities depend on the choice of the {h, k, l}-reflection and give the factor in how many spatial orientations the same lattice distances appear relative to the crystal orientation. Of course, the higher the multiplicity is, the more randomly orientated crystals will contribute. Therefore higher indexed lattices are favored to be used for d-measurements, not only due to their higher angle given lower error in 2θ, but also to their higher multiplicity. After integration the total diffracted power is given by equ. 5-42.

$$P = I_0 \left(\frac{e^4}{m^2 c^4} \right) \frac{V \lambda^3 m F_T^2}{4 v_a^2} \left(\frac{1 + \cos^2 2\theta}{2 \sin \theta} \right) \qquad (5\text{-}42)$$

with diffracted angle θ.

The volume of the effective crystalline material in the sample V is defined as V=M·N·v_a with number of crystals M of a lateral expansion N in unit cell volumes v_a. The structure factor (equ. 5-17) defines the {h, k, l} reflexes weighted by the Lorentz-polarisation factor $(1+\cos^2 2\theta)/2\sin 2\theta$.

A polycrystalline material with big grains in the magnitude of the gauge volume produces segmented Debye-Scherrer cones with bright spots, where big grains accidently appear in Bragg condition. Good diffraction data can be achieved from grain sizes << gauge volume with ~ 1000 grains minimum contributing to the diffracted intensity. Only in some cases, when relative changes in lattice spacing are of main interest, a lower amount of grains can be sufficient. This weak statistic situation should always be handled with care and avoided, if possible. Another intensity fluctuation appears around the Debye Scherrer cone originating from texture. That means that sufficient grains are implemented in the gauge volume, but unnatural strong and weak reflections can be distinguished in different orientations due to favored crystallographic orientations. These effects are problematic as well but more predictable due to a periodicity in intensity fluctuations around the diffracted cone. Texture effects can be reduced by an improved sample orientation relative to the experimental geometry.

In some cases of nano-grained materials, if grain size goes down to the scale of the wave length of the diffracted beam, the peak intensity gets blurred again. The reason is the scaling theorem of the fourier transformed from small crystals. The smaller the periodic regions in real space (crystallites), the stronger the enlargement of the diffraction spot around the Bragg peak will be. Therefore a reduction of diffraction data quality can be expected as well, if the grain size gets too small.

The same assumptions are to be done regarding the white beam diffraction setup with EDS-detectors at fixed 2θ positions. In this case the detector displays a radial section through the Debye Scherrer rings in one Bragg-condition for different wave vectors k. The energy scales equals the d-spacing in the angle dispersive situation. The 1D acquisition reduces the grains orientation variations contributing to the acquired image. Therefore coarse grained structures and texture have significantly higher and unpredictable influence on the diffraction data revealed from energy dispersive setups.

5.5. Stress analysis by diffraction methods

Stresses in materials compose from several components of normal and shearing stresses to a 3D stress state (Fig. 5.20) [18, 19, 20]. These stresses are the components of the stress tensor (equ. 5-43).

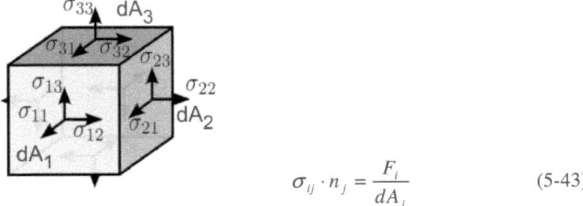

$$\sigma_{ij} \cdot n_j = \frac{F_i}{dA_j} \quad (5\text{-}43)$$

Fig 5.20.: The stress components in a solid [21].

The stress tensor σ_{ij} is tensor of 2nd order between the force F_i (per area A_j) and face normal vector n_j of 1st order. The interatomic spaces in crystalline materials are influenced by elastic deformations described in the strain tensor ε_{ij} (equ. 5-44).

$$\varepsilon_{ij} = \begin{pmatrix} \varepsilon_{11} & \varepsilon_{12} & \varepsilon_{13} \\ \varepsilon_{12} & \varepsilon_{22} & \varepsilon_{23} \\ \varepsilon_{13} & \varepsilon_{23} & \varepsilon_{33} \end{pmatrix} \quad (5\text{-}44)$$

The stress strain relation is given by the Hooks law for a 3D stress state in equ. 5-45.

$$\sigma_{ij} = C_{ijkl} \varepsilon_{kl} \quad (5\text{-}45)$$

The elasticity tensor C_{ijkl} includes the elastic properties of the material such as the Young's modulus and Poisson's ratio.

For cubic systems as investigated in this work the tensor simplifies to only three independent elastic constants C_{11}, C_{12} and C_{44} to (equ. 5-46).

$$C_{ijkl} = \begin{pmatrix} C_{11} & C_{12} & C_{12} & 0 & 0 & 0 \\ C_{12} & C_{11} & C_{12} & 0 & 0 & 0 \\ C_{12} & C_{12} & C_{11} & 0 & 0 & 0 \\ 0 & 0 & 0 & C_{44} & 0 & 0 \\ 0 & 0 & 0 & 0 & C_{44} & 0 \\ 0 & 0 & 0 & 0 & 0 & C_{44} \end{pmatrix} \qquad (5\text{-}46)$$

With the isotropic assumption the shear modulus C_{44} gets to equ. 5-47.

$$C_{44} = \frac{C_{11} - C_{12}}{2} \qquad (5\text{-}47)$$

Hooke's law can now be simplified to equ.5-48.

$$\sigma_{ij} = \frac{E\varepsilon_{ij}}{1+\nu} + \frac{\nu E \delta_{ij} \varepsilon_{kk}}{(1+\nu)(1-2\nu)} \qquad (5\text{-}48)$$

This equation describes the stress strain relation for all stress calculations done in this work [18, 19]. The first term gives the prior stress contribution from strains into stress direction. The second term comes from the Poisson's components of the transversal stresses. Uni-, bi-, tri-axial and isotropic (hydrostatic) stresses can be calculated neglecting any shearing contributions.

In case of residual stresses without any externally applied load the equilibrium condition has to be mentioned (equ. 5-49).

$$\int_V \sigma(\vec{x}) = 0 \qquad (5\text{-}49)$$

with space coordinate $\vec{x}$ and volume V.

This important relation is given by the equilibrium law of forces in a solid. The stress equilibrium helps for the interpretation of stress results, especially in micro stress situations, and allows verifying the consistence of the applied method of measurement and calculation.

Three stress types can be distinguished dependent on their range [20]. Often only macro and micro stresses are mentioned, but in reality three different independent stress phenomena are always present, macro (type 1), micro type 2 and 3 stresses (Fig. 5.21).

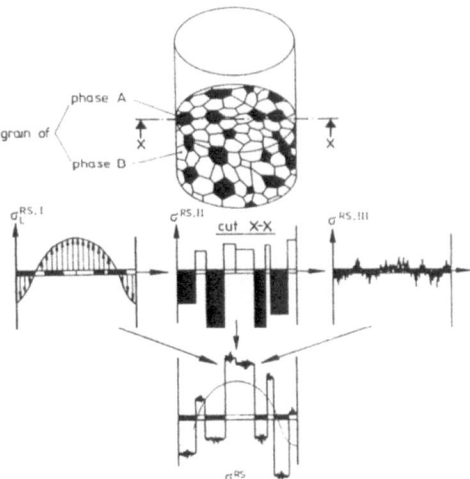

Fig. 5.21.: The residual stress type 1, 2 and 3 in a heterogeneous material [20].

The macro (type 1) stresses are distributed in several grains over large regions of a component with long range linearity. Most of these stresses originate from deformation or thermal gradients during processing. The macro stress components compensate by the equilibrium condition over each of the body cross sections. The stress value obtained by diffractive methods is a phase specific averaged value over the contributing grains in the gauge volume. Therefore the type 1 stresses σ^I are given by equ. 5-50.

$$\sigma^I = \left(\frac{\int \sigma dV}{\int dV} \right)_{several\ crysalls} \tag{5-50}$$

Type 2 micro stresses (σ^{II}) are linear within a grain or big regions of a grain and may vary between several grains (equ. 5-51). These intragranular stresses can be observed in homogeneous as well as heterogeneous materials and are always superimposing the type 1.

$$\sigma^{II} = \left(\frac{\int \sigma dV}{\int dV} \right)_{one\ crysal} - \sigma^I \tag{5-51}$$

Type 3 micro stresses are non-linear short range stresses down to the inter-atomic distance. These stresses generated by dislocations or vacancies are located at a point (x,y) in the crystal and given by equ. 5-52.

$$\sigma^{III} = \left(\sigma - \sigma^{I} - \sigma^{II}\right)_{(x,y)} \qquad (5\text{-}52)$$

The measurable value by (polycrystalline) diffractive methods is always an average over several crystals in the gauge volume, revealing a superposition of all stress types (Fig.5.21) into one value σ^{real} given by equ. 5-53 [22].

$$\sigma^{real} = \sigma^{I}(x,y) + \sigma^{II}(x,y) + \sigma^{III}(x,y) \qquad (5\text{-}53)$$

In this work type 2 micro stresses have been investigated between two phases of a composite material. Diffraction reveals phase sensitive average values of the stresses between the constituents of the material. The structural situation of all the diffraction experiment described in this work is illustrated in Fig. 5.22.

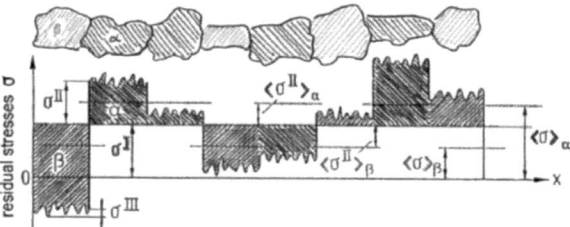

Fig. 5.22.: The dual stress distribution over two phases α and β in a heterogeneous material [22].

The experimental results deliver the phase (α, β) homogeneous stress σ^{II} (averaged) superimposed by pseudo macro stresses σ^{I} in equ. 5-54.

$$\langle\sigma_{\alpha}\rangle = \sigma^{I} + \langle\sigma^{II}\rangle_{\alpha} \quad \text{and} \quad \langle\sigma_{\beta}\rangle = \sigma^{I} + \langle\sigma^{II}\rangle_{\beta} \qquad (5\text{-}54)$$

The volume fraction weighted sum over the stresses in the different phases gives the pseudo macro stress level σ^{I} (equ. 5-55).

$$\sum_{k=1}^{N} f_{k}\langle\sigma\rangle_{k} = \sigma^{I} \qquad (5\text{-}55)$$

The type 2 micro stresses between the phases have to compensate each given in equ. 5-56.

$$\sum_{k=1}^{N} f_{k}\langle\sigma^{II}\rangle_{k} = 0 \qquad (5\text{-}56)$$

The equilibrium condition in this case is fulfilled between both phases weighted by their volume fractions according to the rule of mixture.

The key issue for stress measurements with diffraction techniques is the identification and segmentation of these different stress types. Macroscopic symmetry properties combined with stress equilibrium can be used as boundary condition for segmentation. The gauge volume defined by the experimental setup and radiation capabilities is the collector of the strain data [23]. In general gauge volume << stress range is used to measure the linear lattice displacement Fig. 5.23.

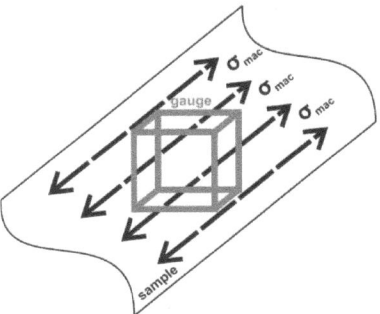

Fig. 5.23.: A gauge volume in an infinitely expanded sample with linear type 1 stresses.

Macro stresses can be directly measured with this method requiring a stress free reference [18, 19]. In case of micro stresses the task is more challenging. Type 2 micro stresses are small in range and inhomogeneous in their magnitude / orientation. These stresses averaged over a gauge volume >> local micro stress range. The strain data will be always a sum over several micro stress components superimposed by the macro stresses [20] (Fig. 5.24).

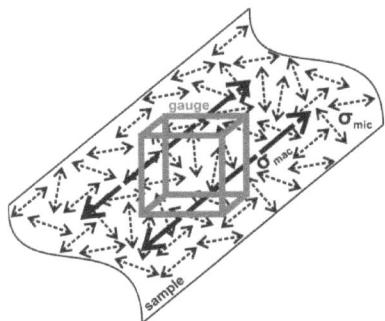

Fig. 5.24.: Infinitely expanded sample with type 1 linear stress state superimposed by type 2, 3 short range stresses < gauge volume.

Fortunately micro stresses develop often between two phases to be resolved by their different crystal structures. In this case a d_0 reference from both components is needed. The capability of high brilliant sources allows the reduction of the gauge volume keeping a feasible acquisition time for detection. Nano spot diffraction (gauge << 0.1 mm^3) can be performed to measure the stresses near interfaces in the bulk of a heterogeneous material. The problem is that a reduction of the gauge volume also goes with a reduction of the grain statistics, if diffraction on a polycrystalline material is performed. The physical limitation of the spatial resolution required for micro stress measurements is therefore limited by the grain size and not only a matter of intensity. Stress equilibrium is not fulfilled, due to the superimposed macro stresses causing a shift of the zero stress value. Another approach of micro stress segmentation can be done by gauge volume > sample (Fig.5.25).

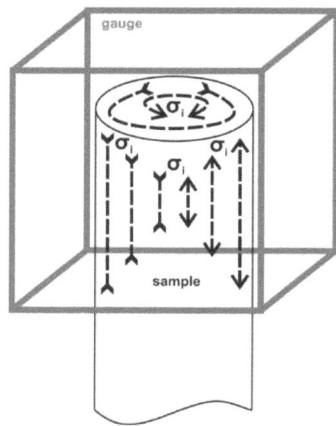

Fig. 5.25.: Gauge volume > sample fulfilling macro stress equilibrium.

In spite of some problems to be discussed later, this method is a powerful tool for interfacial relative stress measurements. The main achievement is elimination of macro stresses by the equilibrium valuable now for the macro stress and the resolving of averaged type 2 micro stresses only. A careful reference measurement has to be performed for the correct d_0 value. With this setup the big gauge volume implements sufficient grains even in coarse grained structures and reduces the acquisition times to a minimum as well which is important for any in situ experiment. The resulting strain data is averaged but allows an isotropic assumption if the stressed region << gauge volume. Therefore the number of required measurement orientations can be reduced according to the system's symmetry axes. With a good sample arrangement single or double shot measurements are sufficient for the complete 3D strain data collection (required for stress calculation) with short acquisition times.

5.6. Diffraction experiment

Laboratory X-ray diffractometers for stress measurements use the Bragg-Brentano geometry (Fig. 5.26) [7]. The X-ray source and detector are arranged around the sample. Both, detector and source can be moved around a circle in an Eulerian cradle, with gauge volume in the circle's center.

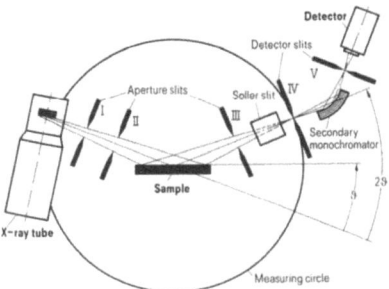

Fig. 5.26.: A laboratory X-ray diffraction setup in an Eulerian cradle [7].

The X-ray tube is fixed, and the sample and the detector rotate in θ and 2θ, respectively. A 1D detector (scintillation counter) is used for angle dispersive peak intensity scans (PSD optional). The monochromatic X-ray beam is directed on the sample defining a surface near gauge volume. The beam penetration of conventional X-ray tubes is limited to several microns (for metals). Therefore "in bulk" measurements are not possible, and some geometrical assumptions have to be made for stress calculation.

Near surface X-ray stress measurements on polycrystalline quasi-isotropic materials base on diffraction line measurements I^{hkl} to evaluate the strain via sample tilting, including the materials Poisson's relation of in a linearly stressed gauge volume [24]. The strains $\varepsilon_{\varphi\psi}^{hkl} = (d_{\varphi\psi}^{hkl} - d_{\varphi\psi}^{hkl})/d_0^{hkl}$ (with d_0^{hkl} as stress free lattice spacing) are evaluated for different azimuthal angle sets (Fig. 5.27).

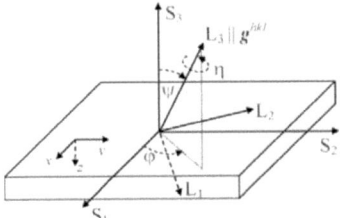

Fig. 5.27.: The coordinate transformation from the sample (true strain directions) into the laboratory system (measureable strain directions) [24].

The diffraction vector g^{hlk} is given relative to the sample reference system S with elastic lattice strain contributions $\varepsilon_{\varphi\psi}^{hkl} = \{\varepsilon_{33}^L\}_{\varphi\psi}^{hkl}$ (L as tensor components relating the laboratory system). The fundamental relations of X-ray stress analysis (XSA) with an averaged mechanical stress $\langle \sigma'_{ij} \rangle$ distributed over several grains in the probed sample volume are given in equ. 5-57.

$$\varepsilon_{\varphi\psi}^{hkl} = \{\varepsilon_{33}^L\}_{\varphi\psi}^{hkl} =$$
$$\frac{1}{2}S_2^{hkl}\sin^2\psi\left[\langle\sigma_{11}^S\rangle\cos^2\varphi+\langle\sigma_{22}^S\rangle\sin^2\varphi+\langle\sigma_{12}^S\rangle\sin(2\varphi)\right]+$$
$$\frac{1}{2}S_2^{hkl}\left[\left(\langle\sigma_{13}^S\rangle\cos\varphi+\langle\sigma_{23}^S\rangle\sin\varphi\right)\sin(2\psi)+\langle\sigma_{33}^S\rangle\cos^2\psi\right]+ \qquad (5\text{-}57)$$
$$S_1^{hkl}\left(\langle\sigma_{11}^S\rangle+\langle\sigma_{22}^S\rangle+\langle\sigma_{33}^S\rangle\right)$$

S_1^{hkl} and S_2^{hkl} are the diffraction elastic constants (DEC) given by equ. 5-58.

$$S_1^{hkl} = \left(\frac{-\nu}{E}\right)^{hkl} \quad \text{and} \quad \frac{1}{2}S_2^{hkl} = \left(\frac{1+\nu}{E}\right)^{hkl} \qquad (5\text{-}58)$$

These DECs can be determined either experimentally or by theoretical calculation of grain interaction models such as Voigt, Reuss, Hill-Neerfeld, Eshelby/Kroener, Vook/Witt.

In case of surface near X-ray diffraction, valid for most laboratory setups, the surface normal components can be neglected. The so called $\sin^2\psi$-method is applicable for stress calculation under such assumptions [18, 19, 20, 23]. Equ. 5-57 simplifies to equ. 5-59 [24].

$$\varepsilon_{\varphi\psi}^{hkl} = \frac{1}{2}S_2^{hkl}\sigma_\varphi\sin^2\psi + S_1^{hkl}\left(\sigma_{11}-\sigma_{22}\right) \qquad (5\text{-}59)$$

The sample is tilted in ψ angles as shown in Fig. 5.28 and the slope of $f(\sin^2\psi)$ is proportional to the in plane stress σ_φ.

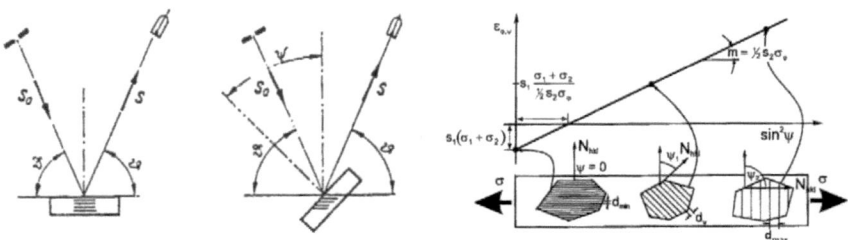

Fig. 5.28.: The sample tilt in the goniometer required for the $\sin^2\psi$-method [25].

The $\sin^2\psi$ method is a powerful tool for laboratory X-ray diffraction with its limitations in penetration depth. The linear approach delivers consistent results without any d_0 reference measurement and is a stable method, if the problem is adequately adapted. Unfortunately the method is strongly sensitive to stress inhomogeneities in the gauge region, such as micro stresses and depth gradients. A deviation from the linear regression will be the consequence from different beam penetration at different ψ tilting angles. Micro stresses inherently cannot be measured with this method, only some limited effects from micro on macro stress levels can be concluded. The acquisition time of one stress scan in a laboratory system comes up to several hours due to > 6 diffraction scans required for one stress measurement and the limited (low) intensity achievable by X-ray tubes. In situ stress measurements with fast acquisition times in a dynamic (especially heterogeneous) stress system are difficult to achieve by this method.

New high brightness synchrotron X-ray sources offer a beam penetration depth increase to several mm even in metals. Direct strain scanning without any interpolation is possible relative to a stress free reference (equ.5-60) [18, 19, 20].

$$\varepsilon = \frac{d - d_0}{d_0}, \quad \varepsilon = \frac{E_0 - E}{E} \quad \text{and} \quad \varepsilon = \frac{\sin\theta_0 - \sin\theta}{\sin\theta} \quad (5\text{-}60)$$

with strain ε, lattice distance d, energy E and diffracted angle θ.

For the stress calculations in this work experimentally determined values of the materials' Young's modulus and the Poisson's relation were used [26]. The temperature dependence of the Young's modulus had to be taken into account for correct stress results at elevated temperatures (thermal cycling experiments). The stresses were calculated with equ. 5-48. A typical beam line setup for synchrotron diffraction is shown in Fig. 5.29.

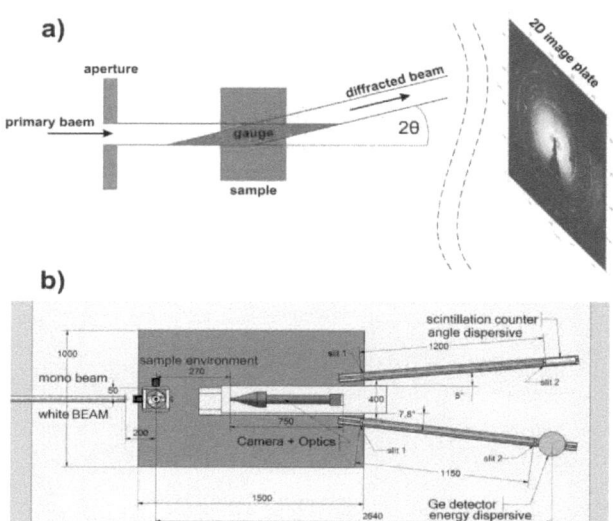

Fig. 5.29.: Synchrotron diffraction setup. Monochromatic angle dispersive in (a), white beam setup in (b). The rhomboidal gauge volume (a) often afflicted with surface effects. An example of a combined diffraction-tomography setup (b).

If monochromatic, a beam at high energies is used for diffraction in transmission through the sample. The energy is manipulated by a double crystal monochromator, adequate to penetration and projection for complete diffraction patterns on the 2D detector plate behind. The Debye Scherrer cones deliver the complete diffraction information. White beam setups are used as well, with one or more EDS-detectors for multi axial stress symmetries. A defined secondary aperture sets the angle of the lattice planes in Bragg-condition. The EDS-spectrum delivers the complete diffraction data of the $\{h, k, l\}$-lattice planes in energy values. The angle usually set to low degrees can be changed for energy spectral scaling. The low angle diffraction condition produces an elongation of the gauge volume into beam direction. The primary aperture limits the parallel beam size on the sample which is rhomboidally enlarged in depth over the sample. High lateral resolutions are possible 2D (x- and z-axis) in the 0.1 mm^3 range with limitations into y-axis. For multidirectional stress measurement the real sample volume contributing the diffracted data has to be handled with care due to incomplete overlapping. A cubic gauge volume would eliminate these geometrical errors.

For deeper penetrations neutron diffraction is applied for strain scanning / mapping in large components. Excellent scattering properties of neutron radiation allow penetration depths of several cm in metals. The diffraction setup can be realized for cubic gauge volumes at angles of ~ 90° (Fig. 5.30).

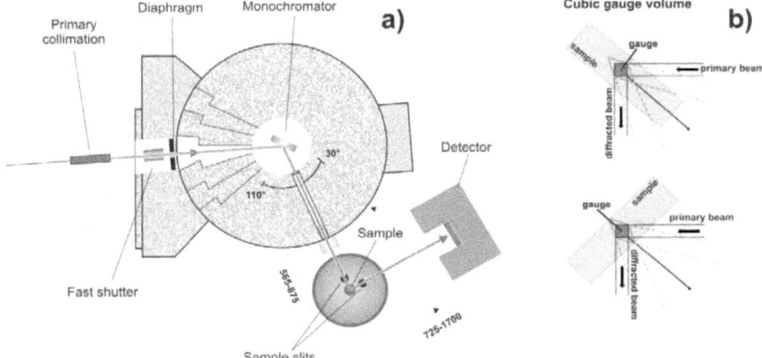

Fig. 5.30.: An angle dispersive neutron strain scanner (Stress Spec FRM2) [16, 17] (a). The advantage of a cubic gauge volume by 100 % overlap of multidirectional scans (b).

The monochromatic beam is focussed by a bent single crystal or crystal array on the sample to increase the intensity. The diffracted beam is scanned by a PSD detector in Bragg condition of the chosen {h, k, l}-reflection of interest. The advantage of a cubic gauge volume allows multidirectional measurements in the same volume element and a reduction of size effects if aligned in the bulk of the sample. A primary and secondary aperture limits the gauge volume in space from ~ 1 mm^3 to ~ 200 mm^3. Lateral scanning over big sample regions is possible.

The pendant to EDS X-ray diffraction is the time of flight (TOF) setup for white beam neutron diffraction. This setup was not used in this work and will therefore not be further discussed.

5.7. Residual stresses in the diffraction pattern

In heterogeneous materials as investigated during this work, the phases can be identified by their characteristic {h, k, l}-reflections. Residual stresses influence the peaks' shapes and positions due to the d-space variations by elastic lattice straining. Micro and macro stresses influence them in a different way dependent on the chosen gauge volume. In the case of type 1 macro stresses a peak shift is induced as shown in Fig. 5.31.

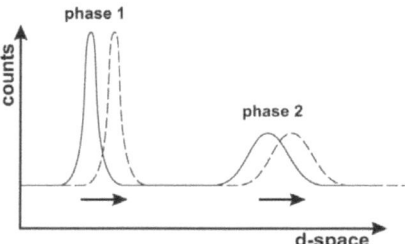

Fig. 5.31.: Peaks shifted by linear macro stresses distributed over both phases.

In this case the gauge volume is small compared to the stress range and a dominating lattice strain component relative to unstrained lattice distance d_0 can be determined. The linear stresses which are distributed over the two phases shift both peaks into the same direction. Different for type 2 micro stresses, which are distributed between the two phases, shifts the peaks relatively against each other (Fig. 5.32).

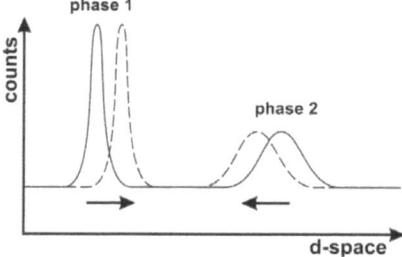

Fig. 5.32.: Peaks shifted by type 2 micro stresses between two phases.

The stress equilibrium is valid for both phases, but not for each on its own. The result is a peak shift of the phases against each other, adequate to the rule of mixture given by their elastic constants and volume fractions. In this case the absolute peak shift is influenced by a stress phenomenon even smaller in range compared to the gauge volume. The different crystal structures divide the effective gauge volume into two overlapping ones, each for every single phase. In case of stressed region < gauge volume with equilibrium inside this region and phase, the peak position remains constant but the widths increase (Fig. 5.33).

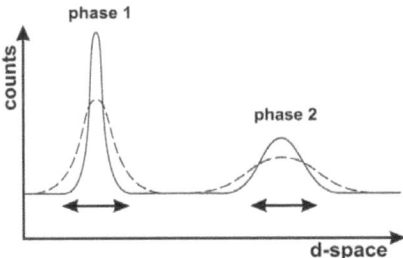

Fig. 5.33.: Peak broadening by stress equilibrium < gauge volume (type 1 and 3 stresses).

Peak broadening due to averaged d-space variation is mostly used to identify type 3 micro stresses with intragranular equilibrium. But if the stressed region < gauge volume condition covers macro stresses at the same time, the peak broadening can be effected by type 1 stresses as well. In case of a sample < gauge volume the macro stress will contribute in peak broadening only and can thus be segmented from the phase sensitive type 2 micro stress state.

5.8. Systematic errors

Diffraction on a polycrystalline sample of finite dimensions and divergent beam conditions causes several geometrical effects on the position and shape of the reflexes. Some of these effects have to be carefully taken into account for stress calculation due to similar geometrical influences on the diffraction pattern like strains have. These effects do not contribute to any error calculation and may be open to misinterpretation of the measured values. Some of these systematic errors are phenomenologically described for monochromatic angle dispersive diffraction in the following passage.

The absorption of the beam plays an important role if the different path lengths in the gauge volume are taken into account (x_1, x_2). In case of gauge volume ~ sample dimension the different regions of the gauge will show entirely different absorption values (Fig. 5.34).

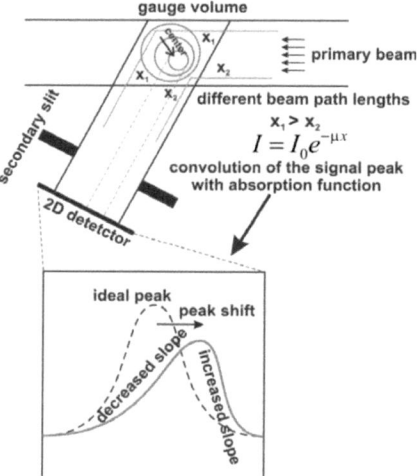

Fig. 5.34.: Gauge volume contributions weighted by beam absorption. The ideal peak is moved towards the source, and the shape is going to be asymmetric.

The absorption gradient over the gauge volume causes a shift of the diffraction center from the center of gravity towards the primary beam dependent on the materials attenuation coefficient μ. A peak shift and asymmetric slopes will be the consequence. If macro stresses in a homogeneous material are of interest, this peak shift can be assumed to be the same for the sample and reference (stress free) and therefore neglected. This will not be the case, if a heterogeneous material of two phases with different attenuation coefficients is investigated. The references will produce an initial peak shift by their different absorption which can be misinterpreted as micro stresses between the phases. The determination of an absolute stress value relative to the components of the material is often difficult.

Surface effects on the peak position can be expected, if the gauge volume is only partially covering the sample and the measurements are affected by the surfaces. This peak shift originates from the geometrical projection of the diffracted volume on the detector (no Fourier translation theorem valuable). Some work was done especially with neutron diffraction in order to study surface effects on strain measurements [27]. The same peak shift phenomenon can be assumed for coarse grained structures where only few single grains contribute in the gauge volume (Fig. 5.35).

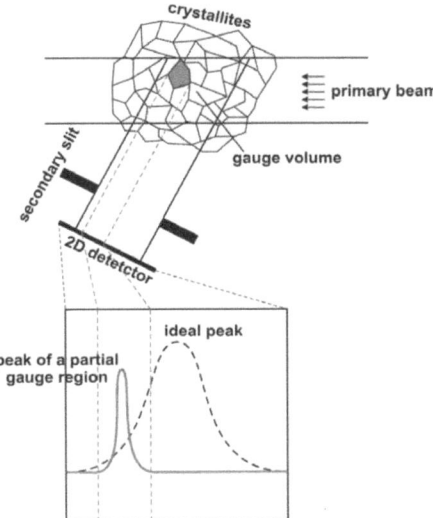

Fig. 5.35.: Coarse grain influence by a limitation of the geometrical gauge to localized diffraction centers. Similar influence if sample < gauge volume.

In this case sample rotation helps to increase the grain statistics and average the artificial peak shift by the 0° and 180° ω-position. Relative stress calculations can be done even without rotation, if the orientation and the position of the crystallites are kept constant relative to the gauge volume during the experiment (thermal cycling, mechanical loading).

If several {h, k, l}-peaks are acquired, a 2θ-angle dependence of the effective gauge volume is the case. Most setups with secondary aperture only (without detector beam collimation) receive the diffraction data from a volume in the sample dependent on the 2θ position (Fig. 5.36).

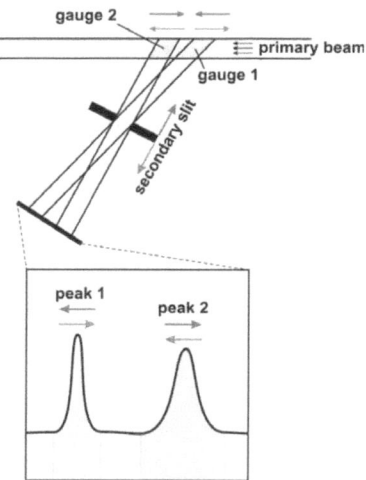

Fig. 5.36.: A virtual peak shift for two peaks at different 2θ angles if the aperture position is changed. Both peaks are contributed by gauge volumes not overlapping in the sample.

The critical issue of such setups is a relative peak shift of two peaks, if the sample-secondary slit position is changed. The result will be a similar phenomenon as caused by micro stresses (Fig. 5.32). Therefore sample and slit position in the diffraction setup have to be kept as untouched as possible within a material-reference scanning procedure.

In energy dispersive diffraction, the geometrical shape of the gauge volume stays constant for all {h, k, l}-peaks due to a fixed 2θ-position. Nevertheless, the acquired reflections in the diffraction patterns are differently weighted over the gauge volume dependent on their energies. In the case of X-ray diffraction (not as systematic as for neutrons) the penetration depth and the materials attenuation coefficient μ (Fig. 5.34, equ. 5-61) depends on the beam's energy. A geometrical shift of the diffraction center varies for each of the {h, k, l}-reflections. In case of high energy X-rays at low 2θ-angles in transmission, this effect is small, but the depth dependence of the spectrum can be important. For reflective measurements at lower energy beam conditions or strong absorbing materials the energy dependent penetration depth cannot be neglected.

5.9. Tomography / Imaging

In this work much emphasis was laid on complementary diffraction techniques using different radiation types for micro stress analysis in composites. The results revealed by tomography mainly base on collaboration with Dr. Guillermo Requena and Dr. Domonkos Tolnai. The experimental methods will be further described in their works [28, 29]. In a brief overview the reader should be given a short introduction into this advanced imaging technique.

5.9.1. Fundamentals

Computed tomography (CT) is a non-destructive imaging technique first applied for medical applications that turned out as powerful tool for materials science over the past decade [30, 31]. Tomography is using individual projections of the beam transmitting the object (radiographies) which are reconstructed to a 3D image. The contrast in absorption contrast imaging is given by equ. 5-61.

$$I = I_0 e^{-\mu x} \qquad (5\text{-}61)$$

The incoming intensity I_0 is exponentially weakened passing through a material at a length of x. The material specific attenuation coefficient μ depending on Z varies considerably for different radiation types (neutrons, electrons, X-rays). Usually X-ray tomography is used for materials science as performed during this work but also neutron and electron tomography gets more relevant in some demanding applications. The basic principle is to rotate the sample in the primary beam (Fig. 5.37) and acquire radiographies in defined angle steps.

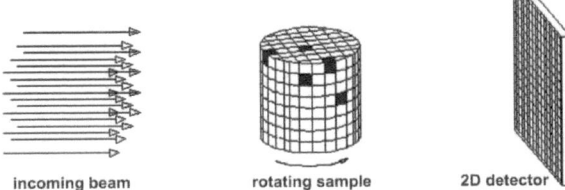

Fig. 5.37.: The principle of parallel beam tomography with a rotating sample in the beam projected onto a 2D detector [32].

The reconstructed 3D data delivers non-destructive spatial information of heterogeneities in solids. In XCT (X-ray computed tomography) more sophisticated contrast giving parameters are used, such as absorption edges, phase contrast or diffraction contrast imaging which is applied if the Z-contrast is not sufficient for structure segmentation. Some examples of experimental setups for XCT as used for this work are presented in the following chapter.

5.9.2. Tomography experiment

In the frame of this work XCT experiments were carried out on FH-Upper Austria, Wels [32]. A sketch of a conventional XCT setup is shown in Fig. 5.38.

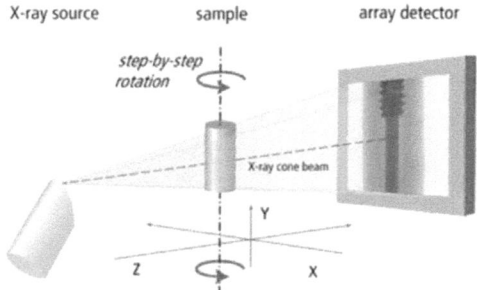

Fig. 5.38.: A cone beam tomography setup with focussing X-ray tube [32].

The focussing X-ray tube produces a point source in the focal point. The divergent polychromatic X-ray beam penetrates the sample, and the radiographies are acquired at the 2D detector behind. The sample size is limiting for transmission, and the pixel size depends on the scanned region (sample area / 1000 ~ voxel size). In a cone beam setup the spot size of the X-ray tube limits physically the possible spatial resolution. Therefore nano-focussing tubes are developed to reach the sub-μm voxel regime. The beam energy critical for penetration cannot be chosen freely (especially for metals), therefore the contrast resolution is often sacrificed to the requirement of penetration with hard X-rays. If polychromatic radiation is used, beam hardening effects may lead to regional contrast weakening. These effects originate from the peculiarity that lower energy radiation is stronger absorbed than high energy X-rays. In XCT typically acquisition times ~ 1h / tomogram are required.

Synchrotron computed tomography was applied in most investigations of this work. These experiments were carried out on ESRF Grenoble at ID15A and ID19 beam lines [33]. A typical white beam diffraction setup is illustrated in Fig. 5.39.

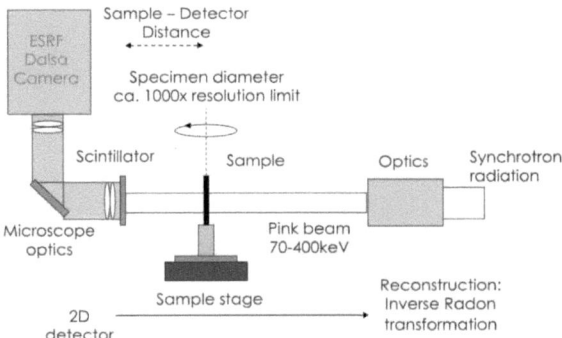

Fig. 5.39.: Synchrotron tomography setup with a parallel polychromatic beam [29].

A parallel beam penetrates the sample and ~ 1000 projections are acquired during one scan with a high resolution CCD behind the sample. The 10^{15} higher brightness reduces the required acquisition time down to ~ 10 sec / tomogram and even shorter. High beam brilliance allows high quality imaging in short time appropriate for in situ tomography. On ID15 diffraction combined with tomography was performed, as described in two of the publications included. The white beam can be replaced by a monochromatic beam as done on ID19. The benefits are no beam hardening effects, and the possibility of phase contrast imaging [29]. With parallel beam the sample to detector distance can be changed which is sensitive to phase contrast required for a more sophisticated method called holo-tomography [28]. This technique allows segmentation of regions with different refractive indices even at equal density. Developed over the past years holo-tomographic imaging was proved to be applicable for materials science on multiphase materials. Unfortunately the physical resolution limit (voxelsize ~ $(0.15 \ \mu m)^3$) of the parallel beam setup is given by the optical wavelength in the CCD. Further resolution increase down to the nm-regime can be achieved by focussing X-ray optics in spite of a parallel X-ray beam. In the work of Dr. Guillermo Requena the Kirkpatrick-Baez optic technique was used for imaging, leading to new insights to microstructural characterization of multiphase materials [34].

5.9.3. Post processing

The reconstruction of the 3D images from the acquired projection bases on filtered back projection algorithms [30]. Not to be explained in detail, this method uses the sinograms with spatial and angle information from all projections during rotation. Fourier transformations and convolution with a filtering function deliver the 3D data in x, y and z direction. The reconstructions were made with great support from the beam line responsible scientists on their sites [32, 33] due to their capacities of large data storage and processing power. After reconstruction the raw data and the reconstructed images were taken for further post processing and analysis.

Image registration and segmentation was done by Wolfgang Altendorfer who used VTK algorithms and special developed software adapted for the task of in situ tomography [35].

References

[1] Skalicky P, Einführung in die Festkörperphysik, vol. 1.3, TU Wien, 1991.
[2] http://www.ww.tu-freiberg.de/mk/Dokumente/Sga_i_2/, may 2011.
[3] Brückel T, Heger G, Richter D, Zorn R, Neutron Scattering, Matter and Materials, vol. 28, Jülich 2005.
[4] Pynn R, Neutron Scattering, Lectures, Los Alamos National Laboratory, 2011.
[5] Murray Gipson J, The Complementary of Real Space and Reciprocal Space, Lecture at Neutron X-Ray School, Oak Ridge National Laboratorys, 2008.
[6] Ewald PP, Fifty Years of X-ray Diffraction, N.V.A. oosthoek's uitgeversmaatschappij utrecht, Netherlands 1962.
[7] Lifshin E, X-ray Characterization of Materials, Wiley-VCH, Weinheim, Germany 1999.
[8] Warren BE, X-ray Diffraction, Addison-Wesley series in metallurgy and materials, 1990.
[9] Behrens M, Powder X-ray and neutron diffraction, Lecture series: Modern Methods in Heterogeneous Catalyst Research, 2011.
[10] Koch E-E, Handbook on Synchrotron Radiation, vol. 1A, North-Holland Publishing Company, Amsterdam 1983.
[11] Mills DM, Synchrotron Radiation Properties and Production, National School for Neutron and X-ray Scattering, University of Chicago, 2008.
[12] http://hasylab.desy.de/, may 2011.
[13] Bonifacio R, Fonda L, Pellegrini C, Undulator Magnets for Synchrotron Radiation and Free Electron Lasers, World Scientific Publishing Co. Pte. Ltd. 1988.
[14] www-xfel.spring8.or.jp, may 2011.
[15] Carpenter JM, Neutron Sources for Materials Research, Tenth international School on Neutron and X-ray Scattering, University of Chicago, 2008.
[16] www.frm2.tum.de, may 2011.
[17] http://www.helmholtz-berlin.de/, may 2011.
[18] Fitzpatrick ME, Lodini A, Analysis of Residual Stress by Diffraction using Neutron and Synchrotron Radiation, Taylor & Francis, London, 2003.
[19] Hauk V. Structureal and Residual Stress Analysis by Nondestructive Methods. Elsevier Science B.V. Sare Burgerhartstraat 25, Amsterdam, NL, 1997.
[20] Macherauch E, Hauk V, Residual Stresses in Science and Technology, vol. 1, DGM Informationsgesellschaft mbH, 1987.
[21] Reimers W, Pyzalla AR, Schreyer A, Clemens H, Neutrons and Synchrotron Radiation in Engineering Materials Science, WILEY-VCH, Weinheim, 2008.
[22] Hauk V, Nikolin NJ, The Evaluation of the Distribution of Residual Stresses of the I. Kind (RS I) and of the II. Kind in Textured Materials, Textures and Microstructures 8&9, 693-716, 1988.
[23] Mittemeijer EJ, Welzel U, Modern Diffraction Methods, Wiley-VCH GmbH, 2011.
[24] Genzel C, Krahmer S, Klaus M, Denks I, Energy-dispersive diffraction stress analysis under laboratory and synchrotron conditions: a comparative study, J. Appl. Cryst. vol. 44, 1-12, 2011.
[25] www.metalle-old.uni-bayreuth.de, may 2011.
[26] Requena G, Creep Behaviour of Discontinuously Reinforced Aluminium Alloys, PhD thesis, Institute for Materials Science and Technology, TU Vienna, 2004.
[27] Fuß T, Wimpory RC, Klaus M, Genzel C, Bridging gaps in surface zone residual stress analysis using complementary probes for strain depth profiling, Mat. Sci. For. Vol. 681, 411-416, 2011.
[28] Requena G, Linking Properties and Architecture in Heterogeneous Lightweight Materials, habilitation, Institute for Materials Science and Technology, TU Vienna, 2010.
[29] Tolnai D, 3D characterization of microstructure evolution of cast AlMgSi alloys by synchrotron tomography, PhD thesis, Institute for Materials Science and Technology, TU Vienna, 2011.
[30] Stock SR, MicroComputed Tomography, Methodology and Applications, Taylor & Francis, USA, 2008.
[31] Baruchel J, Buffiere J-Y, Maire E, Merle P, Peix G, X-Ray Tomography in Materials Science, HERMES Science Publications, Paris, 2000.
[32] http://www.fh-ooe.at/, may 2011.
[33] www.esrf.fr, may 2011.
[34] Requena G, Cloetens P, Altendorfer W, Poletti C, Tolnai D, Warchomicka F, Degischer HP, Sub-micrometer synchrotron tomography of multiphase metals using Kirkpatrick-Baez optics, Scip. Mater. vol. 61, 7, 2009.
[35] Altendorfer W, Void tracking in SiC particle reinforced Al, Diploma Thesis, Institute for Computer Graphics and Algorithms, TU Vienna, Austria, 2008.

6. Discussion of thermal fatigue in MMC

6.1. High volume fraction particle reinforced composites (PRM)

In composites with big CTE mismatch high amounts of micro stresses are generated during temperature change. The constituents are hindered by each other to expand freely and produce thermal stresses. The stresses are distributed between the phases in amplitude indirectly proportional to their volume fraction [1, 2]. The stresses in the bulk volume of a freely expansive composite have to average out according to the rule of mixture as shown in Fig. 2.7. These averaged micro stresses represent a superposition of different contributions dependent on size, shape and volume fraction of reinforcements and on the Young's modulus difference. Several types can be distinguished, as illustrated in Fig. 6.1 as an example for a particle reinforced metal matrix composite with $CTE_r < CTE_m$, $E_r > E_m$:

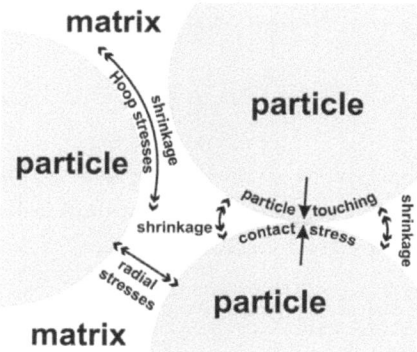

Fig. 6.1.: Micro stresses in a particle reinforced composite after cooling. Tensile Hoop stresses tangential to the particle surfaces are superimposed by radial tension from matrix shrinking between touching particles. Compressive stresses in the particles compensate the matrix tension to equilibrium. Stress peaks are generated in the particles around the touching contact areas where they get pressed onto each other.

The illustration shows the initial stress levels expected in a metal matrix composite with higher particle content produced by liquid infiltration during cooling after production. The matrix percolating the densely packed particle preform is hindered from shrinkage by touching particles. An initial matrix stress level is produced by radial tension between the particles and Hoop tension tangential to the particle surfaces. These short range micro stresses can be superimposed by matrix skin effects or gradients in particle volume fraction. Such mesoscopic effects lead initially to matrix tension near the composite surfaces causing matrix compression between the particles of the particle rich regions.

In the particles a much more complex stress superposition can be expected to compensate the matrix stresses: Compression longitudinal and orthogonal to the interfaces by matrix shrinkage; Particle compression by particle-particle touching in radial direction; Particle tension orthogonal to the surfaces between the particles by matrix hindered in their contraction. The macro stress contribution from the matrix is maintained within the particles in the same way but as well adequate to equilibrium condition averaged over the volume.

During heating the thermal stresses invert from initially overall matrix tension and particle compression into the opposite, respectively. The stress situation at elevated temperatures is shown in Fig. 6.2.

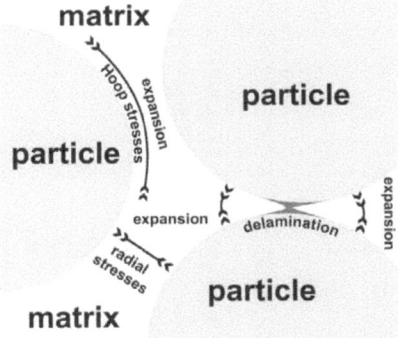

Fig. 6.2.: A particle reinforced composite at elevated temperatures. The expanding matrix between the particles produces Hoop and radial compression. Non wetted particle regions at the previous contact points get opened resulting in delamination and voids.

The matrix produces compressive Hoop stresses near the interfaces due to hindered expansion. If the particles are interconnected or particle movement is blocked by the dense packing of the reinforcement, additional radial matrix compression is built up during heating. Both stress contributions result in a hydrostatic stress situation in the ductile metal matrix at elevated temperatures. The previously mentioned macro-stress contribution may lead to comparably low superimposed compressive matrix stresses in particle poor and surface near regions, compensated by matrix tension in the center- or particle rich-regions.

The stress contributions in the particles are Hoop tension tangential to the interfaces and radial compression from expanding matrix material embedded in between the particles (superimposed by radial particle-particle tension if interconnected). All micro / macro stress contributions are compensated in the different constituents and regions fulfilling the rule of mixture in the volume.

The different kinds of superimposed matrix particle micro stresses are responsible for thermal fatigue load during cycling thermal load. Each of the stress components depends locally on particle shape and reinforcement architecture, which has to be considered according to application requirements.

6.2. Monofilament reinforced composites (MFRM)

In comparison to the (averaged) isotropic stress assumption in particle reinforced composites, a fiber reinforced composite shows a reduced symmetry [2, 3]. A (averaged) two-axial stress system demands separate observation of the stresses longitudinal and transversal to fiber orientation. The longitudinal direction is illustrated in Fig. 6.3 representing the initial condition of an unidirectional fiber reinforced metal matrix composite.

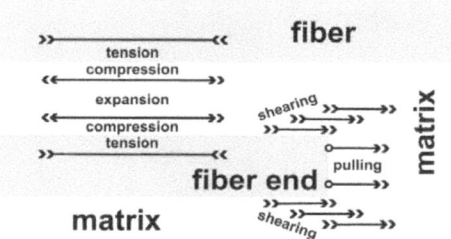

Fig. 6.3.: The micro stress situation in a unidirectional fiber reinforced composite in fiber direction after cooling. Fiber tension and fiber matrix compression is located in the uniformly reinforced region. Near a fiber end shearing and longitudinal tensile stresses in the matrix extended beyond the fiber.

The transverse direction can be assumed as 2-dimensional case of a particle reinforced composite (Fig. 6.1) with fiber cross sections replacing the particles. In diffusion bonded monofilament reinforced composites internal stresses are induced during cooling after their production (similar to sintering or liquid metal infiltration). Initial matrix tension is accommodated by fiber compression fulfilling the rule of mixture. The transversal direction represents a stress controlled situation, influenced by high CTE mismatch stresses in longitudinal direction due to Poisson's contributions. The longitudinal orientation is a strain-controlled geometry with stress amplitudes mainly influenced by CTE mismatch and little by the fiber volume fraction, if interface bonding is strong (elastic assumption). Further macro stresses will superimpose the micro stresses produced by non-uniformities in the fiber volume fraction (surface near regions, badly arranged fibers). During heating, the big CTE mismatch between matrix and fibers will cause high longitudinal micro stresses as shown in Fig. 6.4 (transverse similar to the particles in Fig. 6.1).

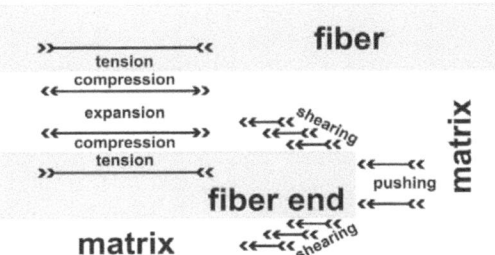

Fig. 6.4.: Micro stresses in a unidirectional fiber reinforced composite at elevated temperatures. Matrix expansion leads to fiber tension and matrix compression in the uniformly reinforced regions. Near fiber ends strong shearing stresses build up and the fibers are pushed by an expanding matrix in between.

The initial stress level will decrease to zero and inverts into matrix compression and fiber tension at high temperatures (Fig. 2.7). The thermal expansion in the strain or stress controlled longitudinal or transversal direction will be reduced in the matrix and increased in the fibers to an intermediate CTE. The overall CTE can be calculated in a first elastic approximation in both directions from the CTEs of the constituents, their volume fractions, and Young's moduli (equ. 2-7). In heat sink materials soft matrix materials are usually reinforced with high strength fibers, providing high fiber aspect ratios. The fiber strength is assumed to be significantly higher than the matrix strength, particularly at high temperatures where the matrix ductility increases. The matrix material has to deform plastically and thus to compensate the big CTE mismatch. The overall CTE (in fiber direction) will diverge from elastic models at elevated temperatures, due to high plastic contributions by displacement of the ductile matrix transverse to the fibers accommodating the high longitudinal stresses.

6.3. Thermal fatigue damage types

In MMCs, high CTE mismatch stresses lead to thermo-mechanical low cycle fatigue loads during thermal cycling if they surpass the matrix yield strength. If 40 % of the homologous solidus temperature of the matrix is surpassed, creep takes place as well. Thermo-cycling creep may occur and cause several damage types, which may affect degradation of the thermal properties of the composite [4]. Plastic deformation of the metal matrix is necessary to compensate the volume difference in a composite with big CTE mismatch and high reinforcement volume fractions during thermal cycling. In Fig. 6.5, a particle reinforced MMC is illustrated with its different damage types. Infiltration voids, originating from production, are not considered as fatigue damage, but reduce the composite's thermal properties in the same way.

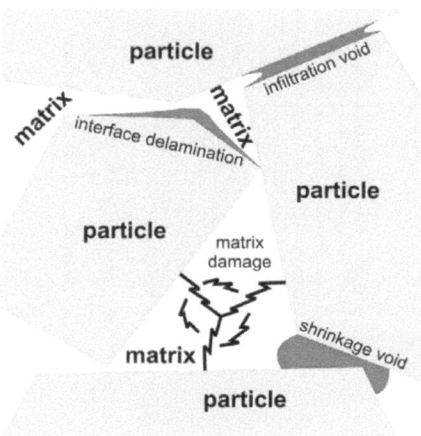

Fig. 6.5.: Thermal fatigue in a particle reinforced composite. Interface delamination, matrix damage, cracks and shrinkage voids can be distinguished.

Voids are formed by matrix shrinking between touching particles during cooling after infiltration even if perfectly infiltrated. Weak bonding strength leads to delamination at the interfaces during thermal cycling. Strong bonding, stronger than matrix strength, leads to interfacial shearing, causing matrix damage like deformation pores. Both damage types degrade the thermal properties during thermal cycling, increasing with the number of cycles and temperature amplitude. In heat sink materials with dominant thermal conductivity of the matrix metal, matrix voids will reduce the thermal conductivity and interface delamination increases the material's volume (and CTE). In composites with high conducting reinforcements, delamination will also reduce thermal conductivity significantly, increasing with increasing temperatures by peeling off the expanding matrix from the stiff reinforcement [5]. Fracture of the reinforcement was neglected due to significant higher reinforcement strength compared to the soft matrix of MMC for heat sink applications.

In fiber reinforced composites, the anisotropic stress situation, dependent on fiber orientation, produces anisotropic thermal fatigue damage mechanisms [6]. A detailed view on the damage types in a uniaxial oriented monofilament reinforced composite is shown in Fig. 6.6.

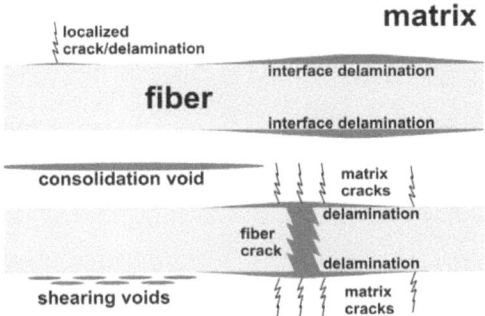

Fig. 6.6.: Thermal fatigue damage in a unidirectional fiber reinforced composite. Interface delamination; Interface shearing voids; Crack near delamination and crack opening; Matrix damage / cracks in crack near regions; Localized delamination around matrix cracks; Consolidation voids (from production, no fatigue damage type).

In transverse direction, Hoop stresses up to matrix yield strength are responsible for matrix deformation and damage during temperature change, if the bonding strength is higher than the matrix shear strength. In a weakly bonded system, matrix expansion will lead to delamination during heating relieving the longitudinal stresses. The bonding increases during cooling by matrix shrinkage onto the stiff fibers at low temperatures (frictional connection). The internal stress level will be shifted into the tensile regime with zero stress at high temperatures. The longitudinal direction shows delamination (weak bonding), matrix damage like shearing voids (strong bonding) and matrix cracking (regional debonding). If initial matrix voids (incomplete consolidation) are present, the radial stress contribution will open them with increasing number of cycles. Fiber cracks cannot be expected to be formed during heating, due to high strength fibers embedded in a relatively soft metal matrix, which is even weaker at high temperatures. If fiber cracks are present, they cause high stress gradients near the ends of the fragments leading to further matrix damage by shearing.

In monofilament reinforced composites, the thermal fatigue damage is a combination of a reduction of the thermal conductivity by matrix damage (if bonding strength higher than matrix strength) and increasing thermal expansion by fiber delamination (if bonding strength weaker than matrix strength). Both will degrade the required thermal properties, demanding a strong matrix sufficiently bonded to the reinforcing fibers.

References

[1] Mortensen A, Kelly A, Zweben C; Comprehensive Composite Materials, vol. 3, Metal Matrix Composites, 541-547, Oxford 2000.
[2] Clyne TW, Withers PJ, An Introduction to Metal Matrix Composites. Cambridge University Press, Trumpington street, Cambridge CB2 1RP, 1993.
[3] You HJ, Bolt H, Overall mechanical properties of fiber-reinforced metal matrix composites for fusion applications, Jour. Nuc. Mat. vol. 305, iss. 1, 14-20, 2002.
[4] Stinchcomb WW, Ashbaugh NE, Composites Materials: Fatigue and Fracture, vol. 4, ASTM, 1916 Race Street, Philadelphia, PA 19103, 1993.
[5] Beffort O, Khalid FH, Weber L, Ruch P, Klotz UE, Meier S, Kleiner S, Interface formation in Al(Si)/diamond composites, Diamond Relat. Mat. 15 (9), 1250, 2006.
[6] Newaz GM, Influence of matrix material on flexural fatigue performance of unidirectional composites, Comp. Sci. Tech. vol. 24, iss. 3, 199-214, 1985.

7. Obtained results

-) The residual stresses in PRM, IPC and MFRM with big CTE mismatch could be evaluated during thermal cycling by methods described above.

-) Infiltration voids in PRM, IPC were identified and their changes in volume fraction were correlated with the temperature cycles by tomography parallel to diffraction. Stress induced visco-plastic matrix deformation was discussed as relevant cause of the macroscopic CTE of the composite.

-) The matrix stress and deformation in PRM, IPC turned out to depend on the different reinforcement architectures which are important for the long term stability of the composite. Si content in an Al matrix results in an interconnected 3D IPC by formation of bridges between the particles. This 3D reinforcement network improves the long term stability under cycling thermal load significantly compared to a system of isolated particles.

-) In MFRM a long range biaxial stress state was identified as combinational stress components, longitudinal and transverse to the fiber direction. The stress amplitudes are directly connected to the interfacial bonding strength.

-) In a soft matrix reinforced by stiff fibers, thermal fatigue damage is mainly located in the matrix or at the interfaces. If stresses exceed bonding strength, delamination will occur producing a CTE increase in fiber direction. If the bonding strength is bigger than the matrix shear strength, shearing voids and cracks in the matrix are formed as soon as stress surpass the matrix shear strength relaxing residual stresses and maintaining the CTE.

-) Fiber cracks which originate from production, will cause accumulating thermal fatigue damage after thermal cycling. The ductile metal matrix suffers severe damage near the fiber ends by shearing and fiber pushing during thermal cycling.

The brief overview of the obtained results will be further discussed in detail in the following publications. Each of them deals with several composite types and heat sink application problems. Metal matrix composites reinforced with high volume fractions of particles or with monofilaments cover the most fundamental types and were investigated by the same methods. Synchrotron as well as neutron diffraction combined with tomography delivered correlating results on stresses and thermal fatigue, influenced by different reinforcement architectures.

Conclusion for MMC design:

Matrix porosity in the range of the volume mismatch between the extreme temperatures expected in service is recommended for IPC and MFRM. This reduces the thermally induced internal stresses. The interface bonding strength needs to be higher than the shear strength of the matrix.

Reinforcement architectures and thermal fatigue in diamond particle-reinforced aluminum

M. Schöbel [a,*], H.P. Degischer [a], S. Vaucher [b], M. Hofmann [c], P. Cloetens [d]

[a] *Institute of Materials Science and Technology, Vienna University of Technology, Karlsplatz 13, A-1040 Vienna, Austria*
[b] *Advanced Materials Processing, EMPA – Swiss Federal Laboratories for Materials Science and Technology, Feuerwerkstrasse 39, CH-3602 Thun, Switzerland*
[c] *Forschungsneutronenquelle Heinz Maier-Leibnitz, Lichtenbergstrasse 1, D-85747 Garching, Germany*
[d] *European Synchrotron Radiation Facility, 6 Rue Jules Horowitz, F-38043 Grenoble, France*

Received 26 May 2010; received in revised form 5 August 2010; accepted 5 August 2010
Available online 16 September 2010

Abstract

Aluminum reinforced by 60 vol.% diamond particles has been investigated as a potential heat sink material for high power electronics. Diamond (CD) is used as reinforcement contributing its high thermal conductivity (TC ≈ 1000 W m K^{-1}) and low coefficient thermal expansion (CTE ≈ 1 ppm K^{-1}). An Al matrix enables shaping and joining of the composite components. Interface bonding is improved by limited carbide formation induced by heat treatment and even more by SiC coating of diamond particles. An AlSi7 matrix forms an interpenetrating composite three-dimensional (3D) network of diamond particles linked by Si bridges percolated by a ductile α-Al matrix. Internal stresses are generated during temperature changes due to the CTE mismatch of the constituents. The stress evolution was determined in situ by neutron diffraction during thermal cycling between room temperature and 350 °C (soldering temperature). Tensile stresses build up in the Al/CD composites: during cooling <100 MPa in a pure Al matrix, but around 200 MPa in the Al in an AlSi7 matrix. Compressive stresses build up in Al during heating of the composite. The stress evolution causes changes in the void volume fraction and interface debonding by visco-plastic deformation of the Al matrix. Thermal fatigue damage has been revealed by high resolution synchrotron tomography. An interconnected diamond-Si 3D network formed with an AlSi7 matrix promises higher stability with respect to cycling temperature exposure.
© 2010 Acta Materialia Inc. Published by Elsevier Ltd. All rights reserved.

Keywords: Particulate reinforced composites; Neutron diffraction; Synchrotron radiation computed tomography; Thermal cycling; Internal stresses

1. Introduction

Materials with high thermal conductivity (TC) in combination with a low coefficient of thermal expansion (CTE) are required for high power electronic devices in order to dissipate the heat from the chips to a heat sink [1]. Pure copper, a high conducting metal with a TC of ~400 W m K^{-1} but a relatively high CTE of ~17 ppm K^{-1}, which is commonly used, has been replaced by particle-reinforced composites such as AlSi7/SiC/70p (AlSiC) (for denotations see Table 1), which exhibit a reduced TC of ~250 W m K^{-1} and a CTE of ~8 ppm K^{-1} [2]. The CTE mismatch between the substrate and the ceramic base plate has to be small to avoid delamination of the solder at the interface due to thermal cycling during use. Increasing power densities demand increased efficiency in the thermal management of high power electronics. Diamond (CD) particles offer a good TC (~1000 W m K^{-1}) and a very low CTE (~1 ppm K^{-1}) [3,4] and could replace the SiC particles in AlSiC [5]. The large CTE mismatch between diamond and a metal matrix (Fig. 1) [6] produces high microstresses at the interfaces between the constituents during changes in temperature (Schöbel, Altendorfer, Degischer, Vaucher,

* Corresponding author. Tel.: +43 15880130836; fax: +43 15880130899.
E-mail address: michaels@mail.tuwien.ac.at (M. Schöbel).

1359-6454/$36.00 © 2010 Acta Materialia Inc. Published by Elsevier Ltd. All rights reserved.
doi:10.1016/j.actamat.2010.08.004

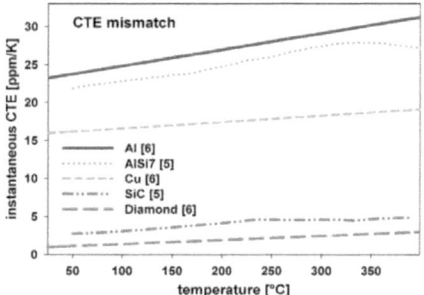

Fig. 1. CTE of Al and AlSi7 matrices, of the SiC and diamond reinforcements and of the composites AlSi7/SiC/70p and AlSi7/CD/60p compared with copper as a heat sink material.

Buslaps, Di Micheli and Hofmann, unpublished work). The use of a Cu matrix was unsuccessful, due to poor bonding with SiC and diamond particles [7]. The weak bonding of diamond with metal matrices needs to be improved to avoid interface delamination during thermal cycling [8]. Al may form interface carbides which improve bonding [4]. Interface reactions can be initiated by heat treatments just below the solidus temperature or by infiltrating SiC-coated diamond particles [9].

Pure Al and AlSi7 matrices have been investigated to determine whether AlSi7/CD/60p composites are stabilized by a three-dimensional (3D) reinforcement network such as in AlSi7/SiC/70p [5]. High particle volume fractions v_p are achieved by melt infiltration of densely packed particle preforms, which are compressed as much as possible by the non-wetting melt so that the particles touch each other [10,11]. The percolating solidified metal within the pores of the preform shrinks during cooling much more than the preform, due to the CTE mismatch between the metal matrix (CTE_m) and the ceramic particles (CTE_p). The volume difference in metal matrix composites (MMC) between the solid matrix v_m and the rigid particle arrangement can be estimated by Eq. (1) with reference to the volume fraction of the matrix [12]:

$$\Delta V(\Delta T) = 3v_m(CTE_m - CTE_p)\Delta T \quad (1)$$

Assuming a densely packed particle preform, the volume mismatch of the matrix with the interstices is determined

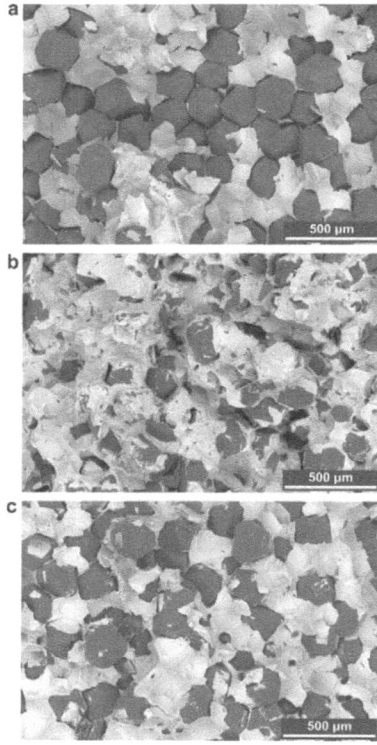

Fig. 2. BSE fractographs of (a) Al/CD/60p exhibiting mainly diamond particles (dark) free of matrix (bright), (b) AlSi7/CD/60p revealing mostly diamond particles covered by Al, and (c) Al/CD$_{SiC}$/60p showing fewer uncovered diamond particles than in (a).

by the volume fraction of the matrix v_m. The volume difference for Al in between 60 vol.% densely packed diamond particles reaches more than 2% during cooling from about 600 °C to room temperature (RT). A temperature change of 320 °C will result in a volume difference of about

Table 1
The diamond particle-reinforced Al composites investigated.

Composite	Matrix	Diamond particles (60 vol.%)	Condition
Al/CD/60p	Al 99.5%	Ø ≈ 200 or 25 μm	As cast
Al/CD/60p ht	Al 99.5%	Ø ≈ 200 or 25 μm	Heat treated, 640 °C, 5 h
Al/CD$_{SiC}$/60p	Al 99.5%	Ø ≈ 200 or 50 μm, SiC coated	As cast
AlSi7/CD/60p	Al + 7%Si	Ø ≈ 200 or 50 μm	As cast
AlSi7/CD$_{SiC}$/60p	Al + 7%Si	Ø ≈ 200 or 50 μm, SiC coated	As cast

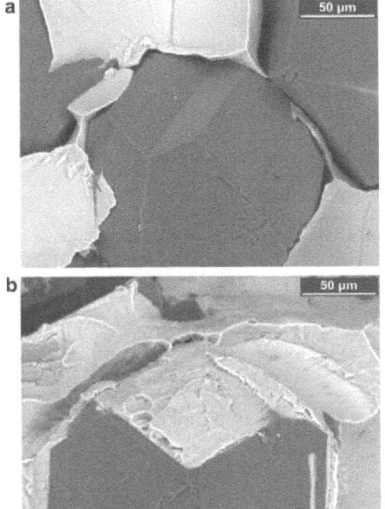

Fig. 3. BSE fractographs of pure Al/CD composites: (a) Al/CD/60p with faces of diamond (dark) and delaminated Al matrix (bright) and (b) Al adhering to SiC-coated diamond particles exhibiting micro-ductile deformation.

1 vol.%. Only a small fraction of deformation (<0.1%) can be accommodated by elastic strain. The resulting microstresses lead to delamination at the interfaces between the particles and the matrix and/or to pore formation within the matrix [12], even if perfectly infiltrated. If such damage was enhanced by thermal cycling it could cause advancing irreversible reduction of the thermal conductivity of the composite.

In a system with strong interface bonding, such as AlSiC, the initial void volume fraction changes during thermal cycling due to stress induced visco-plastic matrix deformation. The resulting void shrinkage produces an anomalous reduction in the CTE of AlSiC above ~200 °C (Schöbel et al., unpublished work). The Si in the AlSi7 matrix forms bridges between the SiC particles during solidification of the eutectic liquid connecting them to a permanent rigid Si–SiC structure [5]. The variation in the volume difference ΔV during heating and cooling is essentially accommodated by the changing void volume fraction. Superior long-term thermal fatigue resistance is expected for an interpenetrating composite compared with reinforcement by isolated particles [13]. Delamination and thermal fatigue damage is more probable in diamond-reinforced Al due to weaker interface bonding and a larger

Fig. 4. BSE images of deep etched MMC showing Si bridges (bright) between the diamond particles (dark): (a) Si in AlSi7/CD/60p produced by squeeze casting at the Swiss Federal Laboratories for Materials Science and Technology [16] and (b) Si lamellae connecting the particles in AlSi11/CD/60p produced by gas pressure infiltration at Plansee Reutte, Austria (http://www.plansee.com).

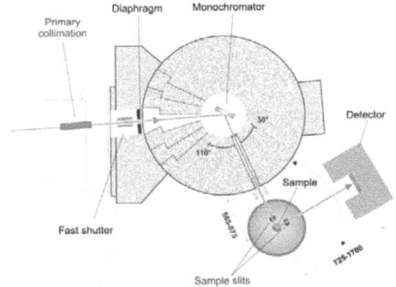

Fig. 5. Sketch of the Stress-Spec instrumental set-up at FRM2 [18]. A Ge single crystal monochromator produces thermal neutrons with a wavelength of $\lambda = 1.67$ Å using the Ge (511) lattice and a flux of 2×10^7 n cm^{-2} s^{-1} on the sample, $2\theta \approx 84.5$–$88°$, acquisition time 3 min.

CTE mismatch than in AlSiC composites. Improvements in interface bonding as well as the role of Si particles during thermal cycling are described here.

Neutron diffraction experiments [14] were performed to measure the microstresses between the matrix and diamond particles in situ during thermal cycling, similarly to previous studies on AlSiC (M. Schöbel et al., unpublished

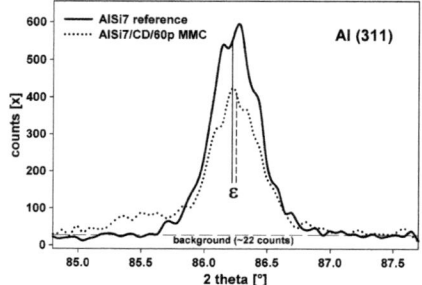

Fig. 6. Al (3 1 1) peak of the AlSi7/CD/60p composite compared with the Al matrix reference indicating the peak quality acquired within 3 min. Matrix strain (ε) in the composite was determined relative to the reference (unreinforced matrix).

work). Correlated in situ high resolution synchrotron tomography was used to visualize plastic deformation of the matrix due to changes in void volume fraction during thermal cycling.

2. Experimental procedure

2.1. Materials description

Diamond-reinforced Al composites were produced by melt infiltration of 60 vol.% monomodal particle preforms by squeeze casting at the Swiss Federal Laboratories for Materials Science and Technology [15]. Approximately 10 mm long cylindrical samples were prepared for neutron and synchrotron experiments with diameters of ~6 and ~0.8 mm, respectively as listed in Table 1. The same experimental set-up was used for the different composite samples to deliver comparable results. Reference measurements for strain determination were made on pure Al and AlSi7 matrix samples with the same dimensions to include the same superimposed macrostresses as in the composite samples and also to reduce diffraction surface effects.

Scanning electron microscope back-scattered electron (BSE) images of fracture surfaces of Al/CD/60p, AlSi7/CD/60p and Al/CD$_{SiC}$/60p (with SiC-coated diamond particles) are shown in Fig. 2. The diamond particles (dark) were partly embedded in the Al matrix (bright). Micro-ductile dimples in the Al indicated good bonding to the SiC-coated diamond particles, as shown in Fig. 3. Si bridges connected the diamond particles in AlSi7/CD/60p, forming a 3D reinforcement network, shown in deep etched AlSi7/CD/60p and AlSi11/CD/60p in Fig. 4, in which the α-Al was leached out. The AlSi7 matrix finally solidified at thermodynamic equilibrium with a 50% AlSi12 eutectic liquid. The grain size of the matrix was in the range 50–500 μm (larger for a pure Al than a AlSi7 matrix). Small diamond particles $\varnothing \sim 25$ and 50 μm) were used for the small tomography samples.

2.2. Neutron diffraction

Neutron diffraction was carried out in a Stress-Spec instrument in the high flux source FRM2, Garching, Germany [16]. A monochromatic neutron beam delivered by a Ge monochromator produces thermal neutrons with a wavelength of $\lambda = 1.67$ Å using the Ge (5 1 1) lattice and a flux of 2×10^7 n cm^{-2} s^{-1} on the sample. The Stress-Spec set-up (Fig. 5) allows strain scans on the Al (3 1 1) lattice planes in the matrix of the composites with an acquisition time of ~3 min in a ~125 mm^3 gauge volume almost completely covering the cross-section of the sample. Unidirectional scans were sufficient, due to the spherical symmetry of the strain system of the particle-reinforced composites (particle size ≪ gauge volume) (M. Schöbel et al., unpublished work). A representative volume is necessary to sample a statistically relevant number of grains contributing to the diffraction peak of the coarse grained matrix. The smaller gauge volumes of synchrotron diffraction would be dominated by crystal texture. The different composites were thermally cycled twice between RT and

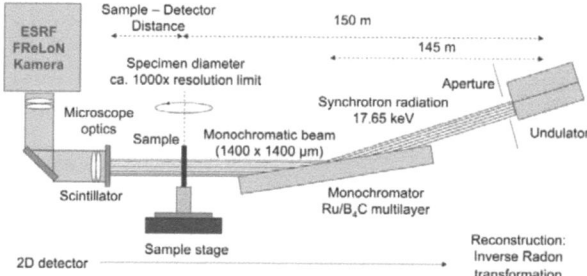

Fig. 7. The tomography set-up of the ID19 beam line at ESRF, Grenoble (http://www.esrf.fr). Parallel monochromatic beam set-up, with 900 images taken during a 180° rotation, CCD camera (2048 × 1024 pixels) in ROI scan mode with 0.3 μm^3 per voxel in binning mode.

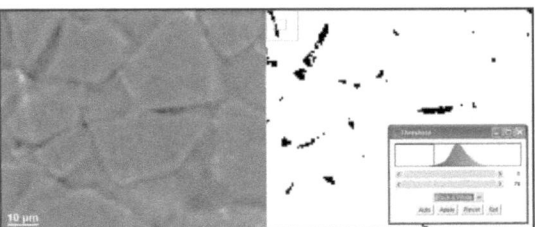

Fig. 8. An image slice from a tomographic scan of AlSi7/CD/60p: (left) reconstructed raw data with dark areas between the diamond particles representing voids and; (right) voids >27 voxels segmented below the threshold value in the histogram.

350 °C (soldering temperature) as in previous synchrotron experiments with the AlSiC composite in ID15A (M. Schöbel et al., unpublished work). The samples were heated by thermal contact with a heating wire and the temperature was measured with a thermocouple mounted on the specimen. Temperature cycles were controlled with a proportional integral derivative controller. The temperature was held constant for ~5 min, longer than the acquisition time (3 min) for isothermal diffraction measurements. The reference measurements on the matrix samples were made under the same conditions using the same set-up. During all experiments the 2D ^{3}He detector (256 × 256 pixels) was set to 82.5° (center position), including Al (3 1 1) at $2\theta \sim 86°$. Vertical summation over the counts gave the peaks relative to the reference (Fig. 6).

The elastic strains were determined in situ during thermal cycling for calculation of the matrix stress as a function of temperature. The gauge volume containing several dia-

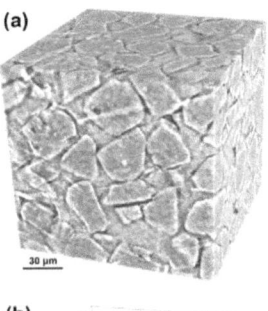

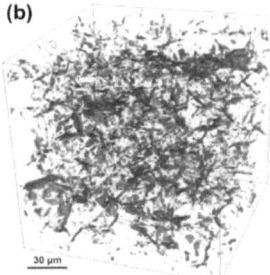

Fig. 9. (a) Microstresses during two thermal cycles of RT to 350 °C, (b) in pure Al matrix with weakly bonded Al/CD/60p and in Al/CD$_{SiC}$/60p with SiC-coated diamond particles, and (c) in α–Al of the AlSi7 matrix with uncoated particles (AlSi7/CD/60p) and SiC-coated particles (AlSi7/CD$_{SiC}$/60p).

Fig. 10. Three-dimension view of a microtomographic image with $(0.6 \mu m)^3$ per voxel of an AlSi7/CD/60p cube of 140 μm^3 under the initial conditions at RT: (a) voids (dark) between the diamond particles mainly at the interfaces and (b) segmentation of voids >5 μm^3 amounting to a volume fraction of 1.6 vol.%.

mond particles embedded in the matrix gave an averaged matrix strain value. The three principal strain orientations ε_i exhibit spherical symmetry with particle sizes smaller than the gauge volume, as shown in a previous work (M. Schöbel et al., unpublished work). Simplification of Eq. (2a) allows calculation of the stress σ from one strain value ε (Fig. 6) according to Eq. (2b) using an isotropic Young's modulus $E(T)$ and Poisson ratio v:

$$\sigma_i = \frac{E(T)\varepsilon_i}{(1+v)} + \frac{E(T)v(\varepsilon_1 + \varepsilon_2 + \varepsilon_3)}{(1+v)(1-2v)} \quad (2a)$$

$$\sigma = \frac{E(T)\varepsilon}{(1-2v)} \quad (2b)$$

These averaged values gave comparable matrix microstress results for the different composites for the same temperature variation, affected by the interface bonding quality and reinforcement architecture. Error values were produced by statistically fitting combined errors from the composite and reference measurements generated by the peak fit software [17].

2.3. Synchrotron tomography

High resolution synchrotron tomography experiments were performed in the ID19 tomography beam line at ESRF, Grenoble, France (http://www.esrf.fr), firstly to visualize the initial void volume fraction at the interfaces between the particles and the matrix and secondly to determine the changes in volume fraction in situ during thermal cycling. A small furnace set-up was used to place the sample as near as possible to the camera optics in order to reduce the phase contrast contribution. A coaxial thermocable heated the sample holder. The temperature was controlled by a thermocouple mounted on the sample at the lower end of the gauge volume. The tomograms were acquired in situ during thermal cycling over the same temperature steps and intervals as for the neutron diffraction experiment to deliver corresponding results.

The monochromatic beam travelled through the sample and was recorded by a high resolution charge-coupled device (CCD) camera behind the sample (Fig. 7). Nine hundred absorption contrast images (2048 × 1024 pixels) were acquired during rotation of the sample through 180° in the beam for each tomographic image. Region of interest scans (0.6 × 0.6 × 0.3mm) in the center of the samples (Ø ~ 0.8mm) delivered high resolution images. Absorption contrast was preferred for void segmentation. The phase contrast of the diamond surfaces was reduced to suppress superimposition on the void absorption contrast, but still show the outline of the particles. The binning mode was

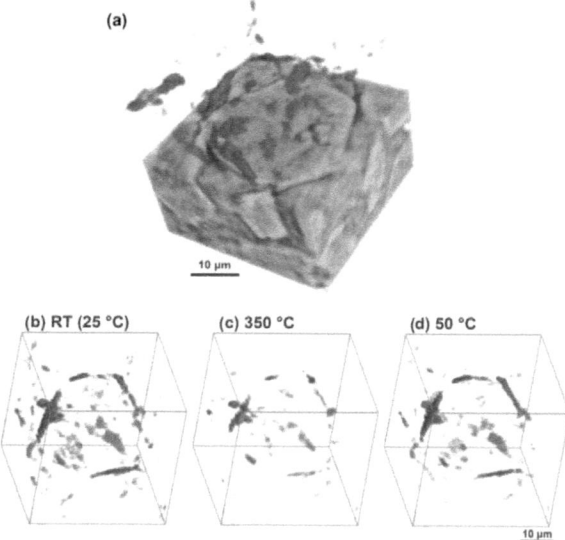

Fig. 11. Three-dimension view of microtomographic images with 0.6 μm³ per voxel of an AlSi7/CD/60p cube of 40 μm³: (a) diamond particles in the lower half of the cube and dark voids in the upper half under the initial conditions at RT, (b) segmented voids in the same region at RT, (c) the same voids segmented at 350 °C, and (d) at RT after cooling again during the first thermal cycle.

applied to the recording at 0.3 μm³ per voxel, yielding a reconstructed voxel size of 0.6 μm³ corresponding to the sum of four pixels, allowing a short acquisition time of ~3 min for each tomographic scan. A contrast resolution of 3 × 3 × 3 voxels provides a void detectability of >5 μm³ (M. Schöbel et al., unpublished work) or detectability of debonding gaps for delamination of >2 μm.

Image motion and ring artefacts were corrected before image reconstruction. The 32-bits 3D volumes were reduced to 8 bits by using only a defined gray value interval centered on the mean value of the 32-bits histogram. This avoids systematic segmentation errors due to brightness drifts between different scans. Void segmentation was carried out by setting a threshold value which was used for the image analysis of all scans in the same way (Fig. 8). Void volume segmentation was performed by voxel counting below the defined gray value, giving quantitative void volume fractions (M. Schöbel et al., unpublished work) only of those voids larger than the resolution limit (~5 μm³). Image registration of the different scans using a registration software tool developed for similar problems [18] allowed region of interest visualization during thermal cycling. Registration was necessary due to sample drift on thermal expansion of the specimen and its mounting. A voxel to voxel correlation for scans recorded during thermal cycling was achieved. The error in void volume fraction of the resolved size class was calculated from the average deviation [18] of the voids segmented using the gray values adjacent to the selected threshold. The slope of the histogram at the chosen threshold gives the corresponding sensitivity of the void volume determination within the segmented volume.

3. Results

Neutron diffraction revealed the internal matrix microstresses in the different composites investigated in situ during thermal cycling between RT and 350 °C. Two cycles were performed to show stress evolution from the beginning, under the initial conditions of the composites at RT, compared with the stresses after the first cycle. Fig. 9 compares the matrix microstress evolution in Al between the composite systems Al/CD/60p, Al/CD$_{SiC}$/60p, AlSi7/CD/60p, and AlSi7/CD$_{SiC}$/60p. There remains some uncertainty about the absolute stress values, including the zero level in the range ±50 MPa (peak offset if sample > gauge volume), with a higher relative accuracy given by the error bars. Compressive stresses built up during heating, reaching a maximum at the peak temperature, when they became extremely high in the case of the AlSi7 matrix. Those stresses decreased during cooling and reverted to tension. The tensile stresses were higher in the Al matrix in which bonding of the diamond particles was enhanced by SiC coating during cooling compared with the uncoated system, but significantly lower than in the AlSi7 matrix composites.

A different matrix stress evolution was observed in the composites with a pure Al matrix (Fig. 9b) and a AlSi7 matrix (Fig. 9c). Initially the AlSi7 matrix stresses were similar to those of the pure Al matrix, but changed during heating to high hydrostatic compression, reaching ~200 MPa at 350 °C in both AlSi7/CD/60p and AlSi7/CD$_{SiC}$/60p. The stresses reverted during cooling to <250 °C, reaching ~200 MPa tension in the Al matrix at RT, explicable by hydrostatic conditioning. A slightly higher matrix stress amplitude was observed in AlSi7/CD$_{SiC}$/60p compared with AlSi7/CD/60p, originating from the higher interface bonding strength of the coated diamond particles.

An AlSi7/CD/60p cube of 140 μm³ with a voxel size of 0.6 μm³ is shown in Fig. 10. Dark voids appear in between the particles, mainly arranged at the interfaces. An initial void volume fraction of ~1.6 vol.% was identified in

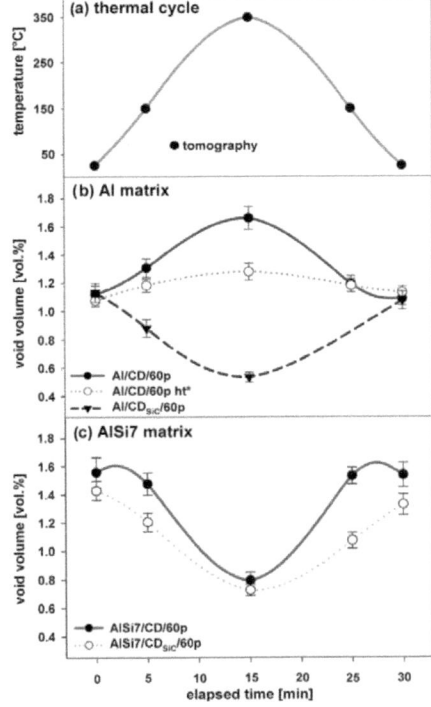

Fig. 12. (a) Quantitative tomographic results of segmented void volume fractions in a 200 μm³ cube during one thermal cycle, (b) in pure Al matrix composites with different interface bonds (*interface reaction on heat treatment at 640 °C for 5 h), and (c) in AlSi7 matrix composites with SiC-coated and uncoated diamond particles.

AlSi7/CD/60p by the described segmentation at RT. Image registration identified the same region of interest from different scans visualized during thermal cycling. A volume of 40 µm^3 around a single diamond particle within the bulk material is shown in Fig. 11. The voids at the interfaces changed in volume during thermal cycling. In AlSi7/CD/60p the voids reversibly closed during heating and reopened during cooling. A larger volume of ~200 µm^3 was chosen for a representative determination of the void volume fraction in Fig. 12. In the pure Al matrix of Al/CD/60p the voids opened during heating and closed during cooling. The voids remained at an almost constant volume fraction in the heat-treated composite (640 °C for 5 h). The change in void volume fraction in Al/CD$_{SiC}$/60p was the inverse of that for Al/CD/60p with an amplitude of 0.6 vol.%: voids shrank during heating and reopened during cooling. This was more pronounced in AlSi7/CD/60p and AlSi7/CD$_{SiC}$/60p, where the voids closed during heating and reopened during cooling with a change in volume fraction of 1 vol.% independent of the SiC coating.

4. Discussion

Voids exist in metal matrix composites produced by melt infiltration of densely packed particle preforms [10], even if perfectly infiltrated. The CTE mismatch between diamond and Al, increasing from 23 to 35 ppm K^{-1} between the ambient and melting temperature of the matrix, produced a volume difference of ~2% for an Al matrix between 60 vol.% particles during cooling from ~600 °C to RT. Microstresses built up in the composite, which could not accommodate the volume difference elastically. Creep

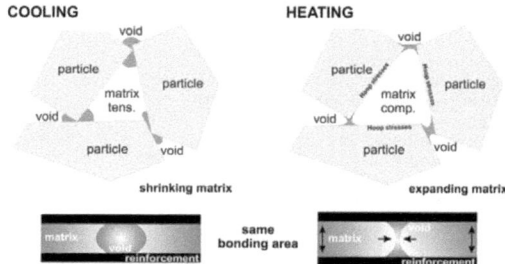

Fig. 13. Isolated particles bonded in a matrix with shrinkage pores in between the particles formed during cooling after infiltration (left). Matrix expansion during heating (right) produced Hoop compression, pressing the soft matrix into the voids. During cooling matrix tension (left) in between the well-bonded touching particles reopened these voids. Changes in void volumes with stable interface contact.

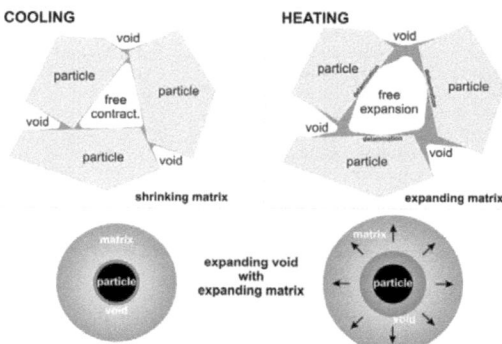

Fig. 14. Isolated particles weakly bonded in a matrix with infiltration voids and regional delaminated interfaces (left). Free matrix expansion was hindered by touching particles but not by weak bonding of particles during heating (right). Voids were opened by expansion of the matrix sponge. During cooling the shrinking Al matrix was pressed onto the particles, closing the voids.

deformation of Al will take place at elevated temperatures and the matrix yield strength may be surpassed approaching RT. Tensile stresses were expected after processing, which may relax during storage. The initial tensile stress level in Al was similar for all the composites investigated. It increased after cooling from 350 °C in the AlSi7 composites.

Figs. 13–15 illustrate the thermal cycling stress situation and the changes in void volume schematically for densely packed diamond particles with and without interface bonding and Si bridges between the particles.

Fig. 13 indicates a stable interface with the pore surfaces changing from concave after cooling to convex during heating [19]. Circumferential Hoop tension built up in the Al close to the interface with the coated diamond particles approaching RT (Al/CD$_{SiC}$/60p in Fig. 9b). The void volume fraction decreased during heating due to the inversion from tensile stress to compression (Al/CD$_{SiC}$/60p in Fig. 12b). Although some delamination at the interfaces was observed, there was strong bonding between the SiC-coated diamond particles and the matrix.

In contrast, the pure Al between uncoated diamond particles exhibited low stress amplitudes (Al/CD/60p in Fig. 9b). A porous sponge-like Al matrix, as shown in Fig. 14 is assumed, with the diamond particles within the pores. The Al sponge expanded during heating, hindered by the stable diamond particles, so that compressive stresses built up. Microtomography revealed an increase in void volume fraction during heating as a consequence of the freely expanding sponge (Al/CD/60p in Fig. 12b). The opposite occurred during cooling, creating tensile stresses, which closed the voids.

The composites containing Si in the matrix alloy, AlSi7/CD/60p and AlSi7/CD$_{SiC}$/60p, formed an interpenetrating structure, with the diamond particles in the matrix connected by Si bridges during solidification of the Al–Si eutectic liquid (Fig. 4). The Si-diamond network formed a rigid open porous cage permeated by α-Al (Fig. 15). The matrix Al shrank during cooling, opening voids to accommodate the volume difference (Fig. 10b). The corresponding hydrostatic tensile stresses relaxed by creep during storage. Only interface bonding with the SiC-coated diamond particles maintained some tensile Hoop stresses. Hydrostatic compression stresses built up during heating due to matrix expansion constrained by the reinforcement network (Fig. 9c). These relatively high multi-axial stresses led to visco-plastic matrix deformation at high temperatures, closing the voids (Figs. 11 and 12c). During cooling from 350 °C stress inversion could be observed at <250 °C, leading to Al matrix tension at RT reopening the voids. The microstress evolution in AlSi7/CD/60p was similar to that in AlSi7/CD$_{SiC}$/60p, with a slightly higher stress amplitude due to the stronger bonding provided by the SiC coating on the diamond particles. Improved bonding by SiC-coated diamonds in the AlSi7 matrix only increased the residual stresses and may not be necessary for stability of such an interpenetrating composite.

However, improved bonding due to SiC coating of the diamond particles in the permeating pure Al matrix produced similar changes in void volume fraction as in the AlSi7/CD/60p samples. Debonding of the particles from the metal matrix would reduce the thermal conductivity of the composite. In the interconnected Si-CD structure of AlSi7/CD$_{SiC}$/60p the dominant compression prevented debonding during heating. The long-term stability may be increased by an interconnected network of reinforcements with interface regeneration at high temperatures, compared with a system of isolated particles. Furthermore,

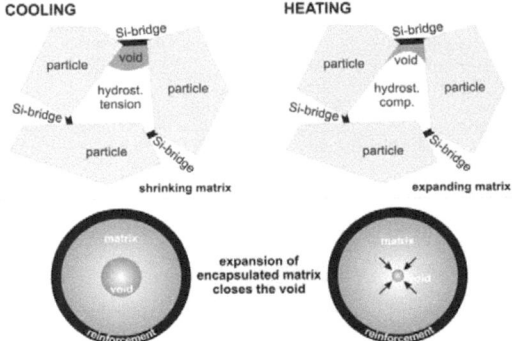

Fig. 15. Matrix embedded in a cage of interconnected reinforcement network. Hydrostatic tension opened shrinkage pores during cooling after infiltration (left). During heating compression of the α-Al within the rigid reinforcement network closed the voids by visco-plastic matrix deformation (right). Matrix tension reopened the voids during cooling (left).

the reduction in void volume fraction during heating would partially accommodate the expanding ductile matrix, leading to a reduction in the macroscopic CTE, as observed for AlSiC (M. Schöbel et al., unpublished work).

5. Conclusions

- Voids form in diamond-reinforced aluminum during cooling from melt infiltration of densely packed diamond particle preforms. Shrinkage of the aluminum matrix between the thermally stable touching diamond particles produces a volume difference of about 2%, which has to be accommodated by delamination and void formation.
- If bonding between particles and matrix is poor, as it usually is between pure Al and uncoated diamond particles, reheating the permeating matrix increases the void volume fraction by free expansion of a ductile matrix foam.
- The bonding can be improved by heat treatment below the solidus temperature, causing the formation of interface Al carbides. Significantly stronger bonding is achieved by infiltration of SiC-coated diamond particles. Tensile Hoop stresses are built up approaching room temperature. The SiC coating causes advantageous void shrinkage during heating.
- Si particles originating from solidification of the interdendritic eutectic liquid, which accounts for 50% of the matrix, connect the diamonds, producing an interpenetrating composite with a 3D network of diamond particles connected by Si between the permeating α-Al. The internal stresses in the Al matrix become compressive during heating and tensile during cooling, creating favorable void closing at elevated temperatures (M. Schöbel et al., unpublished work). If there were no voids, the composite would be likely to break up during heating owing to the volume mismatch.

Thermal fatigue damage occurs particularly during the soldering process, where the electronic packaging is heated to about 350 °C and rapidly cooled. The thermal cycles during service range between −50 °C and +150 °C and might be harmful with increasing number of cycles. An interconnected Si-CD structure in an AlSi7 matrix produces isotropic matrix compression during heating, by which voids and delaminations are closed, so improving conductivity. Tensile stresses induce debonding during cooling by matrix shrinkage, where lower conductivity is required. Thermal fatigue damage propagation may be reduced in AlSi7 matrix systems by creation of an interpenetrating composite, in contrast to the pure Al matrix system with densely packed isolated particles, in which the voids grow with increasing temperature.

Acknowledgements

The research was financed by the ExtreMat 6th framework EU project, and the authors would like to thank all the cooperating partners. Special thanks are due to Manjusha Battabyal for cooperation in the sample preparation. Last but not least, we thank the staffs of FRM2 Garching, Germany, and ESRF Grenoble, France, for their support in helping with, organizing and carrying out the experiments on their sites.

References

[1] Zweben C. J Metal 1992;44:15–23.
[2] Lefranc G, Degischer HP, Sommer HK, Mitic G. In: Massard T, editor. ICCM12 proceedings, ICCM association, Paris; 1999.
[3] Battabyal M, Beffort O, Kleiner S, Vaucher S, Rohr L. Diam Relat Mater 2008;17:1438–42.
[4] Kleiner S, Khalid FA, Ruch PW, Meier S, Beffort O. Sci Mater 2006;55(4):291.
[5] Huber T, Degischer HP, Lefranc G, Schmitt T. Comp Sci Technol 2006;66:2206–17.
[6] Joseph A, King I. Materials handbook for hybrid microelectronics. Norwood, USA; 1988.
[7] Schubert T, Ciupinski L, Zielinski W, Michalski A, Weißgärber T, Kieback B. Sci Mater 2008;58:263–6.
[8] Beffort O, Khalid FH, Weber L, Ruch P, Klotz UE, Meier S, et al. Relat Mater 2006;15(9):1250.
[9] Leparoux S, Diot C, Dubach A, Vaucher S. Sci Mater 2007;57:595–7.
[10] Mortensen A. In: Kelly A, Zweben C, editors. Metal matrix composites, vol. 3. Oxford (UK): Pergamon Press; 2000. p. 541–7.
[11] Clyne TW. Encyclopedia of materials science and technology. Oxford: Elsevier; 2001.
[12] Degischer HP, Lasagni F, Schöbel M, Huber T, Aly MA. In: International conference on advanced materials, ICAMC proceedings, NIIST, Trivandrum-695019, India; 2007.
[13] Schöbel M, Fiedler G, Degischer HP, Altendorfer W, Vaucher S. Adv Mater Res 2009;59:177–81.
[14] Hofmann M, Skirl S, Pompe W, Rödel J. Acta Mater 1999;47(2):565–77.
[15] Beffort O, Vaucher S, Khalid FA. Diam Relat Mater 2004;13:1834–43.
[16] Hofmann M, Schneider R, Seidl GA, Rebelo-Kornmeier J, Wimpory RC, Garbe U, et al. Physica B 2006:385–6. 1035–7.
[17] Fitzpatrick ME, Lodini A. Analysis of residual stress by diffraction using neutron and synchrotron radiation. London: Taylor & Francis; 2003.
[18] Altendorfer W. Diploma thesis, Institute for Computer Graphics and Algorithms. Austria: Vienna University of Technology; 2008.
[19] Nam TH, Requena G, Degischer HP. Composites Part A 2008;39:856–65.

Composites Science and Technology 71 (2011) 724–733

Contents lists available at ScienceDirect

Composites Science and Technology

journal homepage: www.elsevier.com/locate/compscitech

Internal stresses and voids in SiC particle reinforced aluminum composites for heat sink applications

M. Schöbel [a,*], W. Altendorfer [a], H.P. Degischer [a], S. Vaucher [b], T. Buslaps [c], M. Di Michiel [c], M. Hofmann [d]

[a] *Institute of Materials Science and Technology, Vienna University of Technology, Karlsplatz 13, A-1040 Vienna, Austria*
[b] *Laboratory Materials Technology, EMPA, Feuerwerkstrasse 39, CH-3602 Thun, Switzerland*
[c] *European Synchrotron Radiation Facility, 6 Rue Jules Horowitz, F-38043 Grenoble, France*
[d] *Forschungsneutronenquelle Heinz Maier-Leibnitz, Lichtenbergstrasse 1, G-85747 Garching, Germany*

ARTICLE INFO

Article history:
Received 18 May 2010
Received in revised form 15 December 2010
Accepted 23 January 2011
Available online 2 February 2011

Keywords:
A. Metal-matrix composites
B. Thermal properties
B. Porosity/voids
C. Residual stress
D. Non-destructive testing

ABSTRACT

Metal-matrix composites (MMC) are being developed for power electronic IGBT modules, where the heat generated by the high power densities has to be dissipated from the chips into a heat sink. As a means of increasing long term stability a base plate material is needed with a good thermal conductivity (TC) combined with a low coefficient of thermal expansion (CTE) matching the ceramic insulator. SiC particle reinforced aluminum (AlSiC) offers the high TC of a metal with the low CTE of a ceramic. Internal stresses are generated at the matrix-particle interfaces due to the CTE mismatch between the constituents of the MMC during changing temperatures. Neutron and synchrotron diffraction was performed to evaluate the micro stresses during thermal cycling. The changes in void volume fraction, caused by plastic matrix deformation, are visualized by synchrotron tomography. The silicon content in the matrix connecting the particles to a network of hybrid reinforcement contributes essentially to the long term stability by an interpenetrating composite architecture.

© 2011 Elsevier Ltd. All rights reserved.

1. Introduction

Power electronic devices such as IGBT (Insulated Gate Bipolar Transistor) modules are used as converters in hybrid vehicles or railway traction [1]. High power densities generate heat which has to be transported from the ceramic chips through the ceramic substrate, the solder and the base plate into a heat sink. Conventional base plate materials such as Cu or Al, with a high thermal conductivity (TC), do not fulfill the requirements of a matching coefficient of thermal expansion (CTE) in order to avoid delamination of the solder during temperature changes. For this purpose metal-matrix composites are being developed which combine the high TC of a metal with the low CTE of a ceramic. SiC particle reinforced aluminum (AlSiC) has recently been introduced as base plate material [2]. The long term stability under service conditions of such composites is important and has to be guaranteed. Dense particle packing of SiC particles in AlSiC cause significant micro stresses between the particles and the matrix during temperature changes [3]. This is due to the CTE mismatch between the Al in the matrix (23–30 ppm/K) and SiC (6 ppm/K) which may cause delamination between the MMC constituents. However, voids are already present in gas pressure infiltrated AlSiC samples after processing. The anomalous CTE results [4] are compared with thermo-elastic calculations [5] for an AlSi7Mg/SiC/70p composite as illustrated by a simplified diagram in Fig. 1. The Turner model assumes an interconnected reinforcement that is provided by Si bridges between the particles [4]. The difference above 200 °C was explained by stress induced void closure during heating and reopening after cooling [6]. The CTE(T) curve of the composite was simulated using a 2D elasto-plastic FE model allowing a stress induced reduction of the void volume fraction in an AlSi7Mg/SiC/70p composite [7]. The aim of this work was to determine the internal stresses in AlSiC and to correlate them with the void kinetics and the resulting CTE behavior. Synchrotron and neutron diffraction was performed to determine the in-situ micro stresses occurring during thermal cycling. High resolution in-situ synchrotron tomography is used to investigate delamination and void kinetics producing thermal fatigue damage. Pure Al as well as AlSi7 matrix composites with different particle size distributions were tested and the effects on reinforcement architectures described in [8] could be revealed.

* Corresponding author. Address: Institute of Materials Science and Technology, Vienna University of Technology, Karlsplatz 13, E308, 1040 Vienna, Austria. Tel.: +43 15880130836; fax: +43 15880130899.
 E-mail address: michaels@mail.tuwien.ac.at (M. Schöbel).

0266-3538/$ - see front matter © 2011 Elsevier Ltd. All rights reserved.
doi:10.1016/j.compscitech.2011.01.020

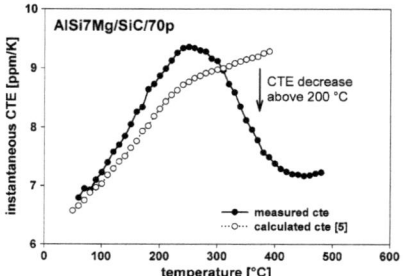

Fig. 1. CTE of AlSi7Mg/SiC/70p between 50 and 500 °C with an anomalous decrease above 200 °C [4] compared to the calculated value [5] for AlSi7 matrix with 70 vol.% SiC particles interconnected.

2. Experimental methods

2.1. Materials description

The same composite was used for dilatometer tests in previous work [4] showing the mentioned anomalous CTE behavior above 200 °C. In order to avoid mechanical finishing after infiltration the Al/SiC/60-70p and AlSi7/SiC/60-70p composites with monomodal ($\varnothing \sim 50\,\mu m$) or bimodal ($\varnothing \sim 5 + 50\,\mu m$) SiC particles were squeeze cast by EMPA [9] in defined dimensions for the neutron and synchrotron experiments as listed in Table 1. Fig. 2 shows the SiC particles and the AlSi12 eutectic between the α-Al dendrites. The matrix of Al/SiC/60p consists of Al grains only, which did not react with SiC owing to the fast infiltration process. The composite without Si in the matrix disintegrates by deep etching and the SiC particles fall apart, but with 7% Si content an interconnected reinforcement network (Si–SiC) remains in shape after removal of the α-Al (Fig. 3). Si bridges connecting the reinforcements to a network were identified similar to those of the AlSi7Mg/SiC/70p composite [4,6] produced by gas pressure infiltration of a trimodal size distribution $\varnothing \sim 3$–$80\,\mu m$ of SiC particles [10] by ELECTROVAC/Austria. Mg has been added to provide some precipitation hardening of the matrix, which does not influence the reinforcement structure.

2.2. In-situ diffraction experiments

Strain measurements were made using synchrotron radiation travelling through the composite samples during thermal cycling [11,12]. The measurements on AlSi7Mg/SiC/70p were performed on ID15A, ESRF, Grenoble [http://www.esrf.fr] using the white beam diffraction setup and two energy dispersive (EDS) Ge-detectors for the two orthogonal reflecting lattice directions relative to the sample [6]. Complete diffraction patterns were acquired within ~2 min simultaneously to micro tomography. The gauge volume ~ 1 mm³ covered the sample diameter ($10 \times 1.5 \times 1.5\,mm^3$) completely by rotating the sample in the beam to increase grain statistics. Diffraction on reference samples of the constituents during thermal cycling provided the $d_0(T)$ necessary for strain calculations. Fig. 4 shows the diffraction pattern of the composite with the corresponding reference scans at RT. The α-Al and Si in the matrix and the SiC particles were identified by their typical lines. Hexagonal contributions (SiC_h) could be identified in the SiC as well but were not taken into account for strain calculation. Some Al lines appear in the SiC particle reference as well due to the powder container made of Al used for diffraction. The marked region of interest (ROI) represents the range in d-space of the compared neutron measurement described in the following.

Strain measurements with an energy dispersive synchrotron setup are more difficult due to the small gauge volume enclosing only a few grains. Therefore neutron diffraction was used for measurements on the coarse grained matrix of Al/SiC/60p and AlSi7/

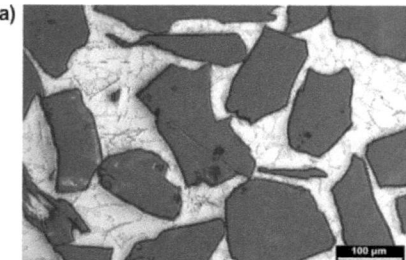

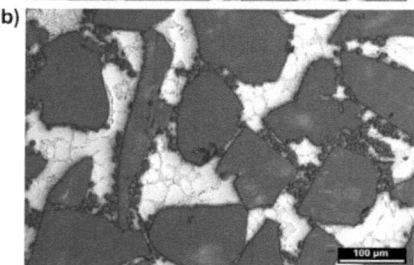

Fig. 2. LOM images of AlSi7/SiC/60p composite (a) monomodal (b) bimodal. Dark SiC particles are embedded in a bright AlSi7 matrix with Si in the interdendritic eutectic (grey).

Table 1
Investigated AlSiC samples.

Composite	Matrix	SiC particle sizes	Vol. frac.	Production	Dimension
AlSi7Mg/SiC/70p	AlSi7Mg	Trimodal $\varnothing \sim 5$ – $80\,\mu m$	70 vol.%	Gas pressure infiltration	$(1.5 \times 1.5 \times 10)\,mm^3$
Al/SiC/60p	Al 99.5%	Monomodal $\varnothing \sim 30\,\mu m$, 100 μm	60 vol.%	Squeeze casting	$d \sim 0.8$ mm, $l \sim 10$ mm for synchrotron tomography and diffraction
Al/SiC/70p	Al 99.5%	Bimodal $\varnothing \sim 5 + 100$	70 vol.%		
AlSi7/SiC/60p	AlSi7	Monomodal $\varnothing \sim 30\,\mu m$, 100 μm	60 vol.%		
AlSi7/SiC/70p	AlSi7	Bimodal $\varnothing \sim 5 + 100$	70 vol.%		$d \sim 6$ mm, $l \sim 10$ mm for neutron diffraction

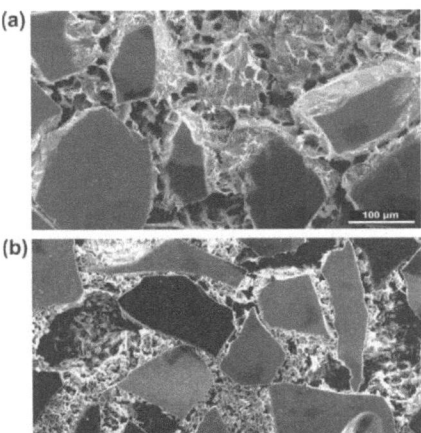

Fig. 3. SEM images of deep etched AlSi7/SiC/60p (a) monomodal SiC (b) bimodal SiC composites as shown in Fig. 2. Si connects the particles to a network of reinforcements.

SiC/60p. Samples were produced for neutron diffraction in cylindrical shape (d ∼ 6 mm, h ∼ 10 mm). Neutron experiments were carried out on the Stress Spec instrument at FRM2 Garching [http://www.frm2.tum.de], [13]. A monochromatic neutron beam ($\lambda = 1.67$ Å) travelling through the sample generates the {h k l} specific diffraction cones acquired by a 2D ^{3}He position sensitive detector (PSD) system. Acquisition times ∼ 3 min were sufficient due to the large gauge volume of $5 \times 5 \times 5$ mm^3 and high monochromatic neutron flux of $\sim 2 \times 10^7 n\,cm^{-2}\,s^{-1}$ on the sample. Fig. 5 shows the Debye Scherrer cones of the observed {h k l} reflections of Al {3 1 1} and of SiC$_c$ {3 1 1} from the polycrystalline sample. A segment of the diffracted cones is projected on the 2D PSD plate. The dominating crystal orientation in the coarse grained matrix produces vertical intensity fluctuations in the Al {3 1 1} diffraction cone. Diffraction patterns of selected peaks of the components were acquired during thermal cycling of the composites. The example of the summed intensities along the Debye Scherrer cones recorded at RT is presented in Fig. 6.

The strains were determined by in-situ neutron diffraction of the composite and its constituents [6] comparable to the synchrotron diffraction but in a smaller d-space range (ROI shown in Fig. 4). Exactly the same setup and sample dimensions were taken for the MMC, the matrix reference and the particle powder container. The samples' diameters were flooded completely by the incoming beam of the synchrotron and also by the neutrons in order to average out macro strains in the samples during thermal cycling and size effects on peak shape. The temperature gradients of the furnace setup for composite as well as for reference samples were neglected.

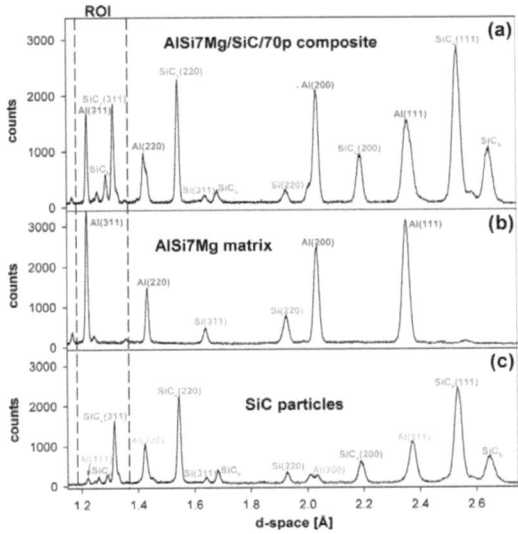

Fig. 4. EDS spectra of synchrotron diffraction at RT of the composite (a) compared with reference measurements of the matrix (b) and the reinforcement particles (c) The ROI represents the d-space range investigated by neutron diffraction in Fig. 6.

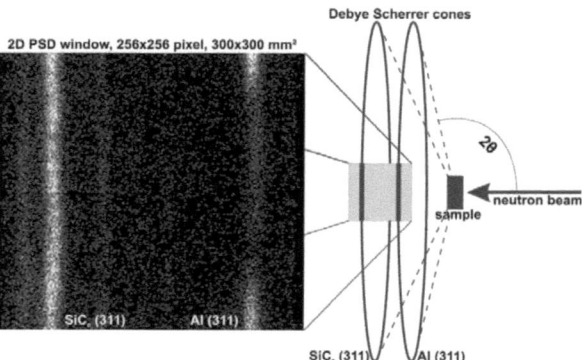

Fig. 5. The neutron diffraction geometry with Debye Scherrer cones behind the sample. The 2D ³He PSD detector on stress Spec, FRM2 shows segments of the projected cones.

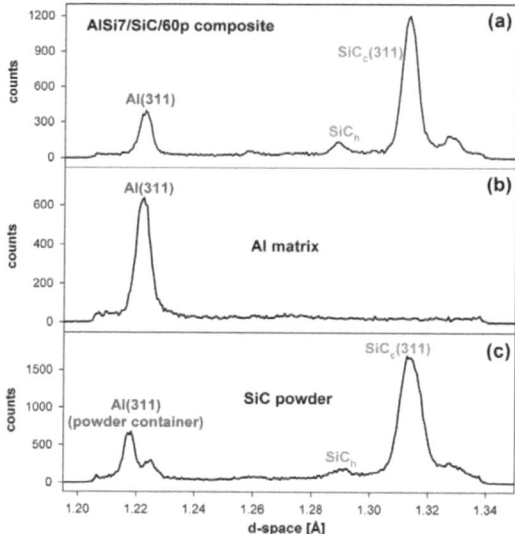

Fig. 6. Vertical sum over the 2D detector image delivers selected {h k l} neutron reflections at a fixed 2θ window of the composite and a reference. The composite (a) compared to Al matrix (b) and SiC powder (c).

Strain calculation was made from synchrotron and neutron diffraction data in the same way. Fig. 7 shows temperature dependence of d-spacing of two dominant matrix peaks Al {2 0 0} and Al {3 1 1} in the composite and the AlSi7Mg reference which were taken from the synchrotron data. The slightly larger d-value in the composite's matrix is reduced during heating ending up smaller than in the matrix reference above 200 °C. The Al {3 1 1} reflection shows a strain hysteresis in the composite due to contributions of plastically deformed matrix material. The hysteresis of the reference in Al {3 1 1} due to interfacial strains between the α-Al and

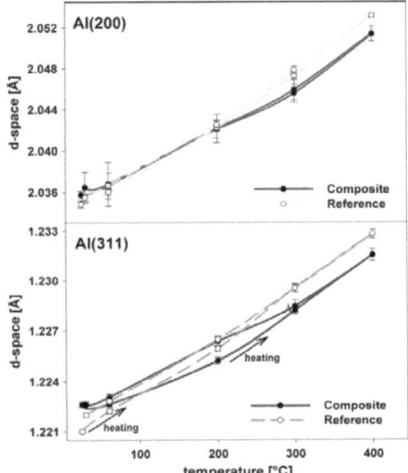

Fig. 7. Changes in matrix d-space during thermal cycling (RT – 400 °C) in the Al {2 0 0} and Al {3 1 1} lattices. The composite shows different behavior during heating and cooling representing the thermal strains.

Si in the matrix are included in the reference as well and therefore not taken into account for the results of the composite's stresses. The AlSi7 matrix was chosen to eliminate these stresses and the d-space variations owing to increasing solubility of Si during heating. The resulting strain values represent the stresses in the phases of AlSi7 matrix produced by the Si–SiC reinforcement architecture only. Strain calculation in the composites was made relative to reference measurements. The strains were calculated from the synchrotron measurement using Eq. (1a) with the composite's peak energies $e(T)$ in the EDS-spectra relative to the reference $e_0(T)$. Eq. (1b) delivered the strains from neutron diffraction with composite's peak angles $\theta(T)$ in the PSD-spectra relative to the reference $\theta_0(T)$.

$$\varepsilon(T) = \frac{e_0(T) - e(T)}{e(T)} \quad (1a)$$

$$\varepsilon(T) = \frac{\sin\theta_0(T) - \sin\theta(T)}{\sin\theta(T)} \quad (1b)$$

EDS peak's energy $e(T)$; PSD peak's angle $\theta(T)$; Strain $\varepsilon(T)$; Temperature T.

The spherical symmetry of averaged micro strains (gauge volume $\gg$ particles size) in and around the particles simplifies the orthogonal stresses from Eq. (2a) to isotropic ones according to Eq. (2b).

$$\sigma_i = \frac{E(T) \cdot \varepsilon_i(T)}{(1+\upsilon)} + \frac{E(T) \cdot \upsilon \cdot (\varepsilon_1(T) + \varepsilon_2(T) + \varepsilon_3(T))}{(1+\upsilon) \cdot (1-2\upsilon)} \quad (2a)$$

$$\sigma = \frac{E(T) \cdot \varepsilon(T)}{(1-2\upsilon)} \quad (2b)$$

Young's modulus E; Poisson's ratio υ; Strain ε; Stress σ; Temperature T.

An isotropic assumption is supposed by the result of synchrotron diffraction in two principal stress directions as shown in Fig. 8, which will be discussed later. Neutron diffraction geometry using unidirectional scans was decided to be sufficient for micro strain determination.

2.3. Fast synchrotron tomography

Synchrotron tomography offers the possibility to investigate thermal fatigue damage in metal-matrix composites [14]. High resolution as well as a short acquisition time on ID15A at ESRF Grenoble enabled in-situ scans during thermal cycling simultaneously to the diffraction analysis. A white beam was used producing absorption contrast images of the sample acquired by a high resolution CCD camera behind the sample. The ~1000 projections of the sample during rotation (180°) were reconstructed to a 3D volume with a voxel size of ~$(1.4\,\mu m)^3$. The acquisition time for one complete tomography scan using the white beam was ~10 min.

Tomography on ID19 at ESRF Grenoble delivered a more accurate absolute value of void volume fractions due to higher resolution. Scans comparable to neutron diffraction on Stress Spec at FRM2 Garching were made with the same composites and conditions, but different samples' shapes. AlSiC has been infiltrated in $10 \times 0.8 \times 0.8$ mm for tomography to achieve a voxel size of $(0.6\,\mu m)^3$ in a region of interest scan of a $0.6 \times 0.6 \times 0.3$ mm^3 volume. A parallel monochromatic beam setup with low energies ($\sim$ 15 keV) was used for good absorption contrast of matrix, particles and voids. The samples were rotated 180° in the beam and the projections were reconstructed to a 3D image similar to the experiments on ID15A. The detector set at high resolution was operated in binning mode to reduce the acquisition time of one scan down to 3 min, which is the same as for neutron diffraction.

The contrasts of the voids and cracks in the matrix were quantified by a feasible threshold value for segmentation in the RT tomogram. The same threshold value was delivered by the 'getAutoThreshold' function of the ImageJ software tool [http://rsbweb.nih.gov/ij]. The same histogram threshold was used for segmentation of voids in all images to investigate relative changes in void volume fractions. It is assumed that the relative changes are not affected by image resolution or by reconstruction artefacts. The absolute values determined are to be considered with respect to the same visibility criterion. The detectability of voids is limited to $\emptyset > 3$ voxel, i.e. 4 μm and 2 μm for ID15A and ID19 tomograms, respectively. The samples' drift between scans due to thermal expansion of the furnace setup was corrected by 3D image registration [15] to investigate regions of interests in the micro tomography scans. Visualization of changes in volume and shape of single voids during thermal cycling can be visualized by voxel to voxel correlation in the different scans after 3D translation and rotation correction of the images.

3. Results

Energy dispersive synchrotron X-ray diffraction delivered diffraction patterns including several {h k l} planes simultaneously (Fig. 4). The strain values were determined relative to reference measurements on matrix samples (Fig. 7). The {h k l} specific strains in the matrix (α-Al, Si) and in the SiC particles are shown in Fig. 8. The orientation independent strains prove the assumption of strain isotropy in the gauge volume ($0.3 \times 0.3 \times 1.5$ mm^3) in directions 0° and 90° relative to the sample. The elastic strain amplitude of Si is bigger than that of the surrounding Al of low

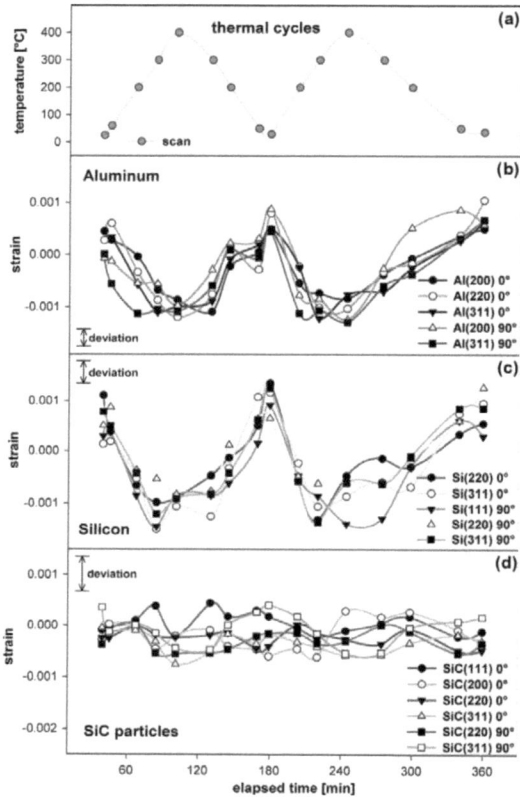

Fig. 8. In-situ strain results (ID15A) during thermal cycling (a) of AlSi7Mg/SiC/70p in (b) matrix Al, (c) matrix Si and (d) SiC reinforcement in orthogonal orientations.

yield strength. The strains in both phases remain more less the same above 300 °C. The SiC particles do not show a significant thermal strain evolution due to their high elastic modulus and high volume fraction of 60–70 vol.%. A more accurate calculation of error (compared to previous results [6]) was achieved by implementation of several peaks and orientations. Matrix micro stresses during thermal cycling are presented in Fig. 9. High hydrostatic compression during heating and tension builds up during cooling within the matrix. The eutectic Si has similar stress behavior as α-Al influenced by the Si–SiC reinforcement. Decreasing stresses can be observed in the Si as soon as 400 °C is surpassed (A), while the stress amplitude in the α-Al remains almost constant >300 °C. The matrix micro stresses appear slightly lower in amplitude after five thermal cycles compared to the first cycle. The same micro stress decrease (A) in the Si can be observed when passing 400 °C after five cycles. The compressive stress in the Al matrix vanishes during cooling (B) inverting into tension at RT, whereas the transition from compression to tension of Si occurs continuously above RT.

Neutron diffraction was made on four composite types with two different matrices and two particle size distributions. Large gauge volumes were necessary for the coarse grained structure of the squeeze cast Al/SiC and AlSi7/SiC with volume fractions of 60 vol.% monomodal and 70 vol.% bimodal SiC particles (Fig. 2). The effect of coarse grained matrix is shown in Fig. 5 by the variations in intensities taken from a cubic volume of $5 \times 5 \times 5$ mm^3. The Al {3 1 1} and SiC$_c$ {3 1 1} peak intensities were summed along the recorded sections of the Debye Scherrer rings. The resulting peak positions were determined at several temperatures, from which the strains were calculated relative to stress free reference measurements. Peak offsets were eliminated by $\Sigma\sigma_i(T) = 0$ over both thermal cycles.

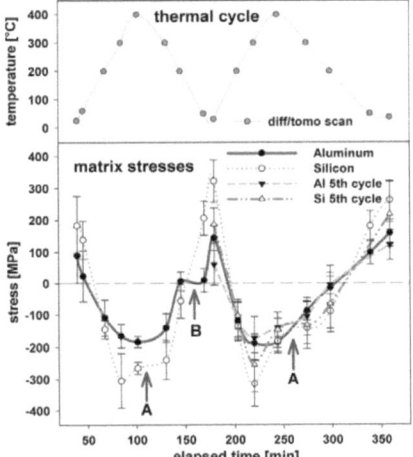

Fig. 9. Micro stresses in the matrix during thermal cycling of AlSi7Mg/SiC/70p calculated from ID15A measurements. Matrix micro stresses in the first 2 cycles (RT – 400 °C) compared to matrix micro stresses after five further cycles. The stress relieve in Si at the beginning of cooling is marked by A. The zero stress level approaching RT is marked by B.

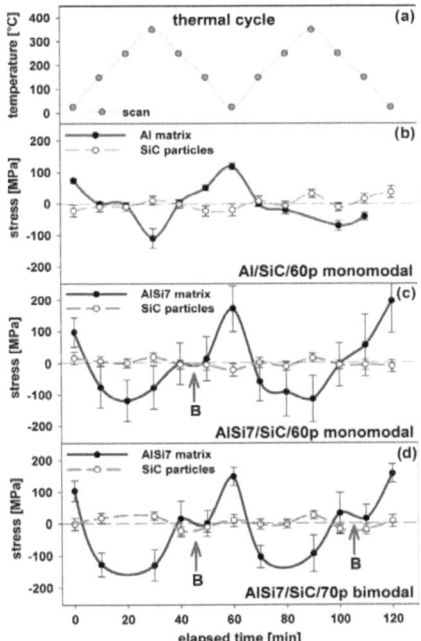

Fig. 10. In-situ neutron diffraction show the micro stresses during thermal cycling (a) in AlSiC: (b) monomodal SiC (Ø ~ 100 μm) with pure Al matrix, (c) with AlSi7 matrix compared with (d) bimodal SiC (Ø ~ 5 + 100 μm) with AlSi7 matrix. The zero stress cooling stage is marked B.

Surface effected peak shifts of the small samples (<gauge volume) could be neglected. The stress calculation was made using Eq. (2b) assuming an isotropic strain state as shown in the synchrotron experiments on AlSi7Mg/SiC/70p (Fig. 8). The micro stresses in the Al-matrix and in SiC were calculated for monomodal and bimodal particle systems (Fig. 10). The error is represented by the peaks' standard deviation [11,12]. Matrix stresses of the Al in AlSi7 are similar to those in Al of the AlSi7Mg matrix from the synchrotron results (Fig. 10c and d vs. 9) with a step of zero stress during cooling (B). The stress evolution in pure Al matrix during thermal cycling (Fig. 10b) can be interpreted only by a similar tendency as of the AlSi7 matrix (Fig. 10c), but with small compression during heating and little tension during cooling. There is no difference in the stress evolution in Al between monomodal and bimodal particle distributions (Fig. 10c and d). The temperature dependence of the neutron diffraction results for AlSi7 with bimodal SiC particle distribution is similar to the synchrotron measurements of the trimodal SiC distribution. A low stress amplitude, independent from the particle distribution, is confirmed for SiC by neutron diffraction.

In-situ synchrotron tomography was performed on ID15A simultaneously to synchrotron diffraction [6] and on ID19 complementary to neutron diffraction (FRM2) [16]. Voids mainly at the interfaces between SiC particles and Al matrix could be visualized by both beam conditions. A change of their volume fraction by stress induced visco–plastic matrix deformation was observed. Individual voids were visualized during thermal cycling after image registration [15] of the high resolution synchrotron tomography scans. A single void between large SiC particles is shown in Fig. 11 during heating and cooling. The volume of a single void cluster (Ø ~ 30 μm) changes comparably to the results taken from a void segmented from an ID15A tomogram shown in Fig. 12c by

about a factor of 5. The void volume fraction of a more representative sample region of ~0.3 mm³ decreases by ~75% from room temperature to 400 °C as shown in Fig. 12b taking into account only voids bigger than 4 μm in diameter due to a voxel size of (1.4 μm)³. Fig. 13 shows a group of voids, which increase 10 times in volume after 25 cycles forming connected voids in between SiC particles. This represents severe thermal fatigue damage at the interfaces.

Synchrotron tomography on ID19 with a parallel monochromatic beam setup and (0.6 μm)³/voxel was made on Al/SiC/60p and AlSi7/SiC/60p (both with monomodal SiC) complementary to neutron diffraction. The volume fraction of voids >1 μm were segmented by in-situ tomographies during the first cycle RT – 350 °C – RT from a representative sample region of (0.3 mm)³. In the AlSi7 matrix the identified void volume fraction decreases from 1.7 vol.% to 0.8 vol.% during heating and increases to 1.3 vol.% after cooling again as shown in Fig. 14. In pure Al matrix, voids decrease from 1.8 vol.% to 1.1 vol.% during heating but do not reopen completely after cooling. The corresponding 3D image views of an Al/SiC/60p cube with the same voids segmented at RT and at 350 °C are shown in Fig. 15. The voids are mainly arranged at the interfaces and shrink as shown in Fig. 14 during heating in

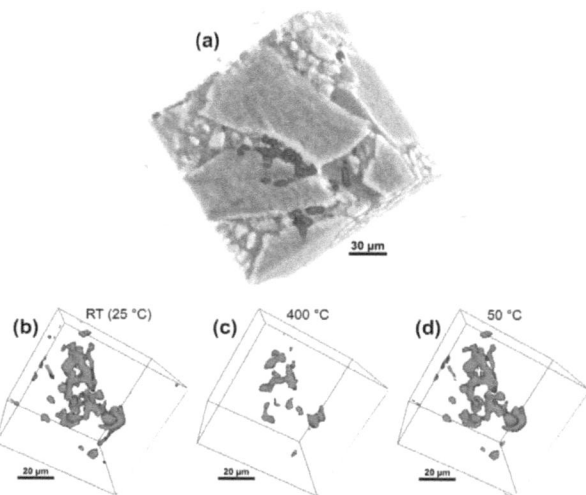

Fig. 11. The voids in AlSi7Mg/SiC/70p located between big SiC particles in trimodal size distribution in an ID15A tomography (1.4 μm)³/voxel with transparent matrix (a), the same group of voids at RT (b), 400 °C (c) and after cooling again close to RT (d).

the pure Al matrix composite due to the big CTE mismatch at the interfaces.

4. Discussion

The big CTE mismatch (ΔCTE) between Al, Si and the reinforcement produces internal stresses in SiC particle reinforced aluminum composites during changing temperatures. The investigated AlSiC samples are produced by infiltration of densely packed particle preforms, which are compressed further by the non-wetting Al melt [10]. Assuming a stress free state at the solidification temperature (neglecting the liquid–solid shrinkage), the matrix in between the rigid SiC particle preform needs to shrink according to the volume difference Eq. (3). (3a) refers to particles embedded in the matrix, whereas (3b) refers to a 3D rigid reinforcement structure as present in AlSi7 matrices where Si forms bridges between densely packed SiC particles. The solidus temperature for AlSi7Mg is 575 °C. The average ΔCTE between Al and SiC amounts to 20 ppm/K, yielding a volume mismatch in Al/SiC/60p according Eq. (3a) of ~2 vol.% when cooled from solidification to room temperature. Applying Eq. (3b) to AlSi7/SiC/60p yields a volume misfit of ~1.3 vol.% during cooling from 575 °C.

$$\Delta V(\Delta T) = 3 v_r \cdot \{CTE_m - CTE_r\} \cdot \Delta T \tag{3a}$$

$$\Delta V(\Delta T) = 3 v_m \cdot \{CTE_m - CTE_r\} \cdot \Delta T \tag{3b}$$

Volume change ΔV, Temperature change ΔT, Volume fraction v. Suffix: Matrix m, Reinforcement r.

Such shrinkage effects cannot be accommodated completely by elastic straining. Therefor 1–2 vol.% voids will be formed by hydrostatic tension in the matrix between the particles, even if infiltration had been perfect. Those voids change their volume during thermal cycling. Visco–plastic matrix deformation results from internal micro stresses close to the matrix yield strength decreasing with increasing temperature. The expanding matrix moves into the voids almost closing them during heating when the particles cannot move and reopens them by tension during cooling again. An initial void volume fraction of >1 vol.% at room temperature in the AlSi7Mg matrix is reduced by 80% at 400 °C by high hydrostatic compression up to ~200 MPa in Al (Figs. 9 and 12). The high hydrostatic micro stresses in Al are explained by the presence of a connected network of Si bridges between the SiC particles [6]. An interpenetrating composite is formed during eutectic solidification of the matrix between the densely packed particles. The Si–SiC network shrinks by 1.5 vol.% less than the Al-matrix in between during cooling and expands by the same amount less during heating. Assuming bonding of the Al, then elastic strains will reduce the volume mismatch. The maximum elastic strains in Al at room temperature were measured to be around 0.1% for an yield strength of ~70 MPa. Since there is no real hydrostatic tension condition in the Al matrix owing to the irregular shape of the reinforcement, at least 1 vol.% of voids can be expected in even perfectly infiltrated AlSi7/SiC/>55p [3] composites at RT. The void volume fraction measured at RT by high resolution synchrotron tomography amounts to >1.2 vol.% of voids >1 μm in diameter (Fig. 14). Thus an almost perfect infiltration quality can be concluded. The lower void volume of 0.15 vol.% measured at ID15A (Fig. 12) with (1.4 μm)³ voxel size refers only to voids >4 μm in diameter. In both cases the void volume fraction is reduced during heating by the inversion of the tensile matrix stress at RT after cooling, to compression at around 200 °C. The ID15A tomograms reveal, that only 20% of the voids >4 μm remain at 400 °C compared to RT. The ID19 tomograms of AlSi7/SiC/60p indicate, that only <50% of the voids >1 μm remain at 350 °C corresponding to a reduction of the void

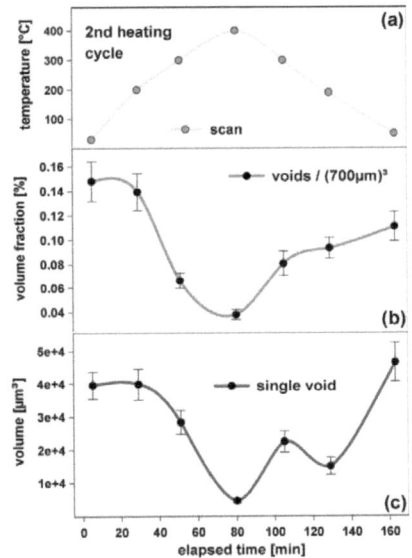

Fig. 12. Results of void segmentation of an ID15A in-situ tomography $(1.4\ \mu m)^3$/voxel of an AlSi7Mg/SiC/70p composite during thermal cycling (a). The voids volume fraction in (b) gauge volume of 0.3 mm^3 is compared with (c) volume changes of a selected group of voids.

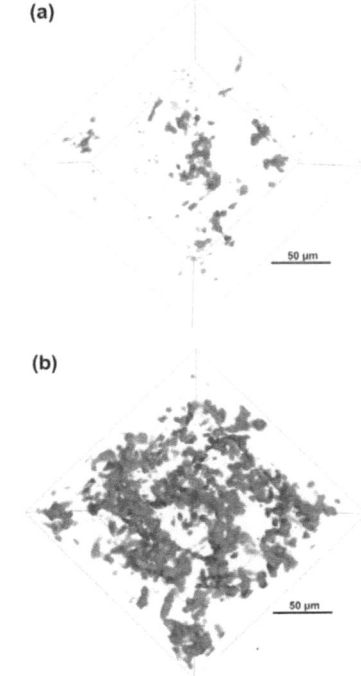

Fig. 13. The same group of voids (Fig.11) in the center of a bigger region $(200\ \mu m)^3$ after the first cycle (RT – 400 °C – RT) (a) and after 25 cycles (b), when the void volume fraction at RT increases 10 times of its initial condition, $(1.4\ \mu m)^3$/voxel, ID15A.

volume fraction from >1.7 vol.% to <0.8 vol.%. High resolution tomograms reveal that the voids close during heating due to visco-plastic matrix deformation by compressive stresses determined by diffraction. This reduction of the void volume fraction, by internal accommodation of the increasing matrix volume, explains the observed anomalous CTE(T) decrease above 200 °C (Fig. 1) [4] which could not be explained by the thermo elastic model [5]. The stress levels in Al and Si undergo a reproducible step in the reduction of compressive stresses >200 °C (marked "A" in Fig. 9) during cooling. The Si bridges between the SiC particles suffer tension as soon as the Al expands in the Si–SiC network. The superposition of compression of Si particles embedded in the Al matrix with the longitudinal tension of the Si bridges between the SiC particles is assumed to be the reason for this step before the stress reduces during cooling. The zero stress level in Al, around 200 °C during cooling ("B" in Figs. 9 and 10) is associated with the reopening of the pores during shrinkage of the Al. Further cooling to RT produces matrix tension reopening the voids more by plastic straining. The 2D unit-cell-model [7] of 70 vol.% SiC connected by 1.6 vol.% Si filled with 28.4 vol.% Al, containing 0.25 vol.% voids at RT, considers only the deformation of the free surface of the Al-matrix along the pores, but does not allow debonding from the SiC particle. The volume of a void in that MMC was calculated to increase by 50% of its initial volume at 500 °C when cooled to RT. When comparing the thermal cycles of Al in the matrix, with and without Si, the internal stresses are much less in the pure Al matrix composite (Fig. 10). The tendency is the same, compression during heating, tension during cooling. The void volume fraction in Al/SiC/60p reduces by 30% during heating. In the pure Al matrix a system of touching particles exists which cannot follow the shrinkage of the matrix. If the interface bonding is strong then movement of particles by the expanding metal matrix during heating is hindered by touching neighboring particles, even when they are not connected by Si bridges [16]. Therefor Eq. (3a) does not represent the situation. The void volume fraction increases with the number of thermal cycles indicating that debonding advances by low cycle fatigue damage mechanisms. After 25 cycles strong thermal fatigue damage can be observed through an increase of the void volume fraction of up to 10 times of the initial value (Fig. 13). After five cycles the micro stress amplitude is reduced in AlSi7Mg/SiC/70p (Fig. 9). The increase in the void volume fraction occurs faster in the Si free Al matrix than in those with Si–SiC networks. Similar micro stresses were observed in AlSi7/SiC/60p with monomodal SiC as well as in AlSi7/SiC/70p with bimodal SiC particle distributions [17] by in-situ neutron diffraction (Fig. 10). The increase of the stress amplitude in Al with increasing SiC volume fraction is within the scatter range. The SiC particle stresses are not detected to compensate the matrix stresses inversely.

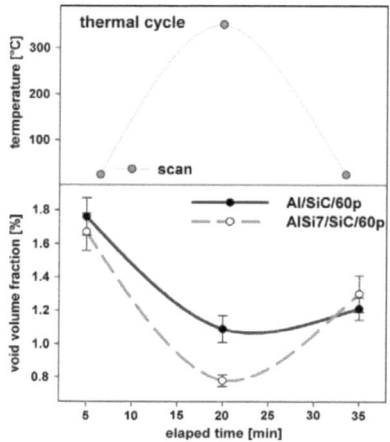

Fig. 14. Void volume fractions during heating and cooling (RT – 350 °C) in monomodal (∅ ~ 50 μm) SiC reinforced Al with and without Si content in the matrix.

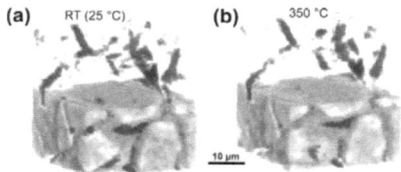

Fig. 15. Al/SiC/60p monomodal at RT (a) at 350 °C (b), (0.6 μm)³/voxel, ID19.

5. Conclusions

- Metal-matrix composites with densely packed ceramic particles contain a certain volume fraction of voids after cooling from the processing temperature. This is due to the large CTE mismatch between the constituents resulting in the elasto-plastic and viscous deformation of the ductile matrix.
- Al matrices with Al–Si eutectic solidify after melt infiltration of the particle preform forming an interpenetrating composite consisting of a network of ceramic particles connected by rigid Si bridges.
- The Al matrix suffers hydrostatic tensile stresses after cooling from processing. During reheating the stress inverts into hydrostatic compression to significantly higher levels within Si–SiC networks than in pure AlSiC composites.
- During reheating the voids close partly by viscous flow of the matrix and reopen plastically during further cooling, parallel to the micro stress oscillations.
- The void kinetics during thermal cycling reduces the CTE at elevated temperatures, where creep of the matrix becomes effective. Voids influence the macroscopic CTE behavior by allowing for plastic deformation of the softer phase.
- Thermal cycles exceeding 150 K imposed on AlSiC represent low cycle fatigue exposure on the Al matrix at the interfaces with the rigid reinforcement producing debonding, which increases with the number of cycles. The damage advances significantly faster in pure Al matrix than in composites with Al–Si eutectic.

Acknowledgements

The research was financed by the ExtreMat 6th framework EU project, and the authors would like to thank all the cooperating partners. Also thanks owing to Mr. A. Rhyswilliams for his language corrections during finalization of the work. Last but not least we would like to thank the helpful support from the staff of ESRF Grenoble France and FRM2 Garching Germany helping organizing and carrying out successful experiments on their sites.

References

[1] Zweben C. Metal-matrix composites for electronic packaging. J Metal 1992;44:15–23.
[2] Lefranc G, Degischer HP, Sommer HK, Mitic G. In: Massard T, editor. ICCM12 proceedings. Paris: Electronic Support;; 1999.
[3] Elomari S, Boukhili R, San Marchi C, Mortensen A, Lloyd DJ. J Mater Sci 1997;32:2131–40.
[4] Huber T, Degischer HP, Lefranc G. Schmitt. Comput Sci Technol 2006;66:2206–17.
[5] Turner P. J Re Nat Bu Stan 1946;36:239–50.
[6] Schöbel M, Requena G, Kaminski H, Degischer HP. Mater Sci For 2008;571-572:413–8.
[7] Nam TH, Requena G, Degischer HP. Composites: Part A 2008;39:856–65.
[8] Schöbel M, Fiedler G, Degischer HP, Altendorfer W, Vaucher S. Adv Mater Res 2009;59:177–81.
[9] Beffort O, Long S, Cayron C, Kuebler J, Buffat PA. Comput Sci Technol 2007;67:737–45.
[10] Michaud VJ, Suresh S, Mortensen A. Fundamentals of metal matrix composites. Boston: Butterworth-Heinemann; 1993.
[11] Fitzpatrick ME, Lodini A. Analysis of residual stress by diffraction using neutron and synchrotron radiation. London: Taylor & Francis; 2003.
[12] Hauk V. Structural and residual stress analysis by nondestructive methods. NL: Elsevier B.V. Amsterdam; 1997.
[13] Hofmann M, Skirl S, Pompe W, Rödel J. Acta Mater 1999;47(2):565–77.
[14] Terzi S, Salvo L, Suéry M, Limodin N, Adrien J, Maire E, et al. Scripta Mater 2009;61:449–52.
[15] Altendorfer W. Void tracking in SiC particle reinforced Al. Diploma Thesis, Institute for computer graphics and algorithms. Austria: TU Vienna; 2008.
[16] Schöbel M, Degischer HP, Vaucher S, Hofmann M, Hofmann M, Cloetens P. Acta Mater 2010;58(19):6421–30.
[17] Molina JM, Narciso J, Weber L, Mortensen A, Louis E. Mater. Sci. Eng. A 2008;480:10–20.

Thermal fatigue damage in monofilament reinforced copper for heat sink applications in divertor elements

M. Schöbel [a,*], J. Jonke [a], H.P. Degischer [a], V. Paffenholz [b], A. Brendel [b], R.C. Wimpory [c], M. Di Michiel [d]

[a] *Institute of Materials Science and Technology, Vienna University of Technology, Karlsplatz 13, A-1040, Austria*
[b] *Max-Plank-Institut für Plasmaphysik, Boltzmannstrasse 2, D-85748 Garching, Germany*
[c] *Helmholtz Zentrum Berlin, Hahn Meitner Platz 1, D-14109 Berlin, Wannsee, Germany*
[d] *European Synchrotron Radiation Facility, 6 Rue Jules Horowitz, F-38043 Grenoble, France*

ARTICLE INFO

Article history:
Received 18 June 2010
Accepted 22 December 2010
Available online 30 December 2010

ABSTRACT

In fusion reactor systems extreme conditions require materials with high temperature and radiation resistance. The divertor component consists of a plasma facing W plate attached to a Cu heat sink to extract the heat from the nuclear reaction chamber into the cooling medium. The Coefficient of Thermal Expansion (CTE) mismatch between the W plate and the Cu heat sink causes interface delamination reducing the long term stability of the divertor.

To avert this problem, composites are developed as interlayer materials with a high thermal conducting Cu matrix reinforced with up to 50 vol.% SiC or W monofilaments to increase the mechanical strength and to reduce the CTE mismatch. Thermal stresses are transferred from the macroscopic interface between the components into the bulk of the composite. Oscillating micro stresses may lead to fiber delamination and matrix damage during thermal cycling. Different matrix alloys, fiber materials and interface designs are investigated.

In situ neutron diffraction performed during thermal cycling show the effect of bonding strength on the stress amplitudes expected under service conditions. The long term stability is tested by measurements after further ex situ cycling. Thermal fatigue damage and its propagation are visualized by in situ as well as ex situ high resolution synchrotron tomography. The combination of both methods helps to understand the strain induced damage mechanisms. Weak bonding leads to delamination of the fiber–matrix interfaces. Strong bonding causes severe matrix deformation and damage. Fiber cracks originating from sample production cause accumulating thermal fatigue damage during thermal cycling.

© 2011 Elsevier B.V. All rights reserved.

1. Introduction

The extreme conditions in fusion reactor systems put high demands on materials. The challenges for material science in such high temperature and irradiation conditions are complex. Heat produced during the thermonuclear reaction has to be transported from the reaction chamber into a cooling medium [1]. One of the highest temperature loads appears in the divertor elements (20 MW/m^2). The divertor collects the residual from the thermonuclear reaction out of the reaction chamber and transports the heat into the coolant (H$_2$O, He or LiPb). A plasma facing W plate is attached to a CuCrZr heat sink. Pulsed operated reactor types [2] produce high interfacial stresses between the Cu and the W plate during operation due to the Coefficient of Thermal Expansion (CTE) mismatch (ΔCTE $\sim$ 12 ppm/K). Delamination at the interface would reduce the cooling efficiency of the divertor. Monofilament reinforced copper is developed to be applied as interlayer material between the two components. The high Thermal Conductivity (TC) of the copper matrix reinforced by fibers with a low CTE combines the good thermal properties of both [3] providing in addition high temperature strength as well. Micro stresses are generated in the composite between the metal matrix and the reinforcement during changing temperatures [4]. The bonding between the fibers and the matrix has to withstand the stress created in service conditions therefore SiC and W monofilaments in Cu [5] and CuCrZr matrices [6,7] with different interface designs [8,9] have been investigated. Tensile tests, acoustic emission and heat flux tests were performed to correlate the interface bonding strength with thermal properties of the composites [10].

The higher the bonding strength the higher the thermal micro stresses between the fibers and the matrix will be, due to the fiber–matrix CTE mismatch. Weak bonding leads to delamination at the interfaces and fiber sliding during thermal cycling. If the

* Corresponding author. Address: Institute of Materials Science and Technology, Vienna University of Technology, Karlsplatz 13, E308, 1040 Vienna, Austria. Tel.: +43 15880130836; fax: +43 15880130899.
E-mail address: michaels@mail.tuwien.ac.at (M. Schöbel).

0022-3115/$ - see front matter © 2011 Elsevier B.V. All rights reserved.
doi:10.1016/j.jnucmat.2010.12.242

bonding strength is higher than the shear strength of the matrix, severe matrix deformation lead to matrix and/or fiber damage [3]. The magnitude of the thermally activated micro stresses in the fiber and the surrounding matrix are strain controlled longitudinal to fiber orientation [11]. The matrix strain fields around each fiber overlap in the composites of fiber volume fractions considered here. Relaxation will produce opposite plastic deformation in longitudinal and transverse direction to the fibers. An intermediate bonding strength (sliding < bonding strength < fiber/matrix damage) fulfills the requirements for a good thermal stability of the composite during thermal cycling.

Goal of this work was to investigate the internal stresses in situ during external thermal load as expected under service conditions. The influence of bonding strength on micro stresses is correlated with thermal fatigue damage. In situ neutron and synchrotron diffraction were made during thermal cycling for phase sensitive micro-stress measurements in the bulk of the composites.

2. Experimental methods

2.1. Materials description

Samples of monofilament reinforced metals (MFRM) with different Cu matrices, fiber materials and coatings were investigated as listed in Table 1. SiC monofilaments of SCS0 and SCS6 type (Ø ∼ 140 μm) with volume fractions from 10 to 30 vol.% and W wires (Ø ∼ 140 μm) with 20–50 vol.% were used as reinforcement (Fig. 1). Pure Cu as well as CuCrZr (Cr ∼ 0.7 wt.%; Zr ∼ 0.1 wt.%) matrices were tested [6–10]. The MFRM samples were produced by Deutsches Zentrum für Luft- und Raumfahrt (DLR) and some by Max Planck Institute (IPP) in cooperation with European Aeronautic Defence and Space Company (EADS) [7,12]. First step in production is the fiber coating with an interlayer followed by Cu or CuCrZr matrix material deposition by magnetron sputtering or galvanic Cu coating in thickness dependent on the required volume fraction. The coated fibers are filled into a preform encapsulated in a matrix tube and compacted by hot isostatic pressing (HIP) at 600–700 °C with 100–120 MPa for Cu and at 750–850 °C with 100–170 MPa for CuCrZr matrices [7]. Samples resulted with a MFRM core of Ø ∼ 3 mm for neutron diffraction and Ø ∼ 1.2 mm for synchrotron diffraction/tomography. Different interface designs were developed to increase fiber–matrix bonding strength: SiC fibers coated with Ti (SCS6) or TiTaC (SCS0) interlayers during magnetron sputtering (IPP) [8]; W wires with graded or etched interfaces [9].

2.2. In situ diffraction experiments

In situ strain measurements were performed in the composites by neutron diffraction during thermal cycling [13]. The experi-

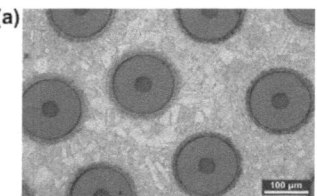

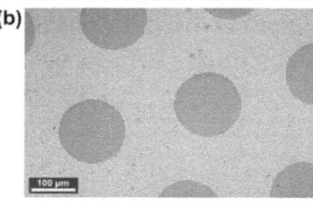

Fig. 1. Light optical image of (a) Cu/SiC/20 m (IPP) reinforced with SiC SCS6 fibers and (b) CuCrZr/W/20 m (DLR) reinforced with W wires.

ments were carried out on the E3 strain scanner at HZB Wannsee, Germany (Fig. 2) [http://www.helmholtz-berlin.de, 14] using a monochromatic neutron beam with λ = 1.486 Å and flux ∼ 5 × 10^6 n cm^{-2} s^{-1} on the sample. The Cu {3 1 1} matrix peak was scanned using a position sensitive ^{3}He detector (PSD) with 256 × 256 pixel and 1.38 × 1.38 mm²/pixel. The detector window was set to acquire the W {2 2 0} peak as well. Thermal load was applied with a well contacted heating wire. Temperature cycles were controlled by a thermocouple mounted on the sample in the gauge volume and connected with a computer programmed proportional – integral – derivative controller unit. The sample holder allowed stress free expansion/contraction of the composites with good positioning in the beam with the help of two aligning gaps. In situ measurements were made during thermal cycling parallel to reference scans on Cu matrix bulk samples and W powder at the same temperature steps. The MFRM samples were cut out from tensile test samples (d ∼ 3.5 mm, l ∼ 16 mm, initially) [10]. Diffraction was made in two principal stress directions of the cylindrical samples (Fig. 3). The primary aperture was chosen as big as possible (6 × 6 mm²) covering the complete sample diameter. The secondary slit was set to 2 mm producing a gauge volume of 70 mm³ with subsequently 5 min acquisition time. The gauge volume with the shape of a parallelepiped covers a slice through the complete sample diameter appropriate for micro-stress segmentation. The smaller secondary slit and monochromator optics [14] minimized potential aberration effects, as a surface effect. The macro stresses (temperature, stress gradients) superposing micro stresses were eliminated using a matrix reference sample of same size/shape under the same conditions. The resulting relative peak positions were determined to calculate the strains (Fig. 4). Matrix stresses were calculated from bidirectional scans using Eq. (1) simplified for a rotational symmetric stress state to Eqs. (2a) and (2b) [15].

$$\sigma_i = \frac{E(T)\varepsilon_i}{(1+\upsilon)} + \frac{E(T)\upsilon(\varepsilon_1 + \varepsilon_2 + \varepsilon_3)}{(1+\upsilon)(1-2\upsilon)} \quad (1)$$

$$\sigma_{matrix}^{long} = \frac{E_{matrix}(T)\left((1-\upsilon_{matrix})\varepsilon_{matrix}^{long} + 2\upsilon_{matrix}\varepsilon_{matrix}^{trans}\right)}{(1+\upsilon_{matrix})(1-2\upsilon_{matrix})} \quad (2a)$$

Table 1
Investigated MFRM samples.

Composite	Matrix	Fiber	Vol. frac.	Fiber/wire interface
Cu/SiC/20 m^{IPP}	Cu	SiC (SCS6)	20 vol.%	Uncoated
Cu/SiC + Ti/20 m^{IPP}	Cu	SiC (SCS6)	20 vol.%	Ti coating
Cu/SiC + TiTaC/20 m^{IPP}	Cu	SiC (SCS0)	20 vol.%	TiTaC coating
CuCrZr/SiC/10 m^{DLR}	CuCrZr	SiC (SCS6)	10 vol.%	Uncoated
CuCrZr/SiC + Ti/15 m^{DLR}	CuCrZr	SiC (SCS6)	15 vol.%	Ti coating
CuCrZr/SiC + Ti/30 m^{DLR}	CuCrZr	SiC (SCS6)	30 vol.%	Ti coating
Cu/W graded/20 m^{IPP}	Cu	W wire	20 vol.%	Graded
Cu/W etched/20 m^{IPP}	Cu	W wire	20 vol.%	Etched
CuCrZr/W/30 m^{DLR}	CuCrZr	W wire	30 vol.%	Uncoated
CuCrZr/W/50 m^{DLR}	CuCrZr	W wire	50 vol.%	Uncoated

Composite description – matrix/fiber/volume fraction, monofilament^{manufacturer}.

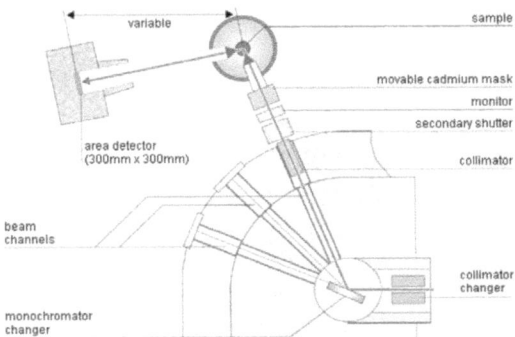

Fig. 2. The experimental setup of the E3 strain scanner at HZB, Wannsee, Germany [13]. $\lambda \sim 1.486$ Å, flux $\sim 5 \times 10^6$ n/cm^{2}s (on the sample), 2D detector ^{3}He, 256 × 256 pixel.

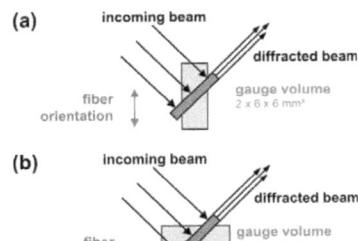

Fig. 3. Diffraction setup for biaxial scans, (a) longitudinal, (b) transversal relative to the cylindrical MFRM samples. Gauge volume covers the diameter completely.

$$\sigma_{fiber}^{long} = \frac{E_{fiber}(T)(1 - \upsilon_{fiber})\varepsilon_{fiber}^{long}}{(1 + \upsilon_{fiber})(1 - 2\upsilon_{fiber})} \quad (2c)$$

where $E(T)$ is the temperature dependent Young's modulus, υ the Poisson ratio, ε the strain and σ the stress.

Emphasis was laid on the longitudinal direction to characterize bonding strength related to the matrix micro-stress. The W wire stresses could not be determined in the transverse direction due to a strong wire texture. A simple uniaxial strain state assumption relative to W powder reference scans was made using Eq. (2c) with ($\varepsilon_{long} = (d - d_0)/d_0$ and $\varepsilon_{trans} = 0$) [16]. The calculations delivered relative results to compare fiber bonding qualities.

Synchrotron diffraction was performed to investigate effects of fragmented fibers on micro stresses during thermal cycling. The experiments were carried out on the high energy beam line ID15A at ESRF Grenoble [http://www.esrf.fr]. A monochromatic X-ray beam ($\lambda = 0.1337$ Å) transmitted the poly crystalline sample forming Debye Scherrer cones in space. The projections of these cones were acquired with a 2D image plate detector behind the sample (Fig. 5) with 2300 × 2300 pixel in 350 × 350 mm^2 and an acquisition time of 10 s with 90 s readout. The small gauge volume produced by a primary aperture of 0.2 × 0.2 mm^2 was increased over the sample diameter of d = 1.6 mm by sample rotation (360°/scan) to a volume of $\sim$0.5 mm^3. Improved grain statistics during scans of 30 s delivered representative diffraction data. The strains longitudinal and transversal to the fiber direction were separated from the 2D image and calculated using a matrix reference. From the relative peak shift, strain could be determined and subsequently stress calculated using Eqs. (2a) and (2b). The SiC fiber

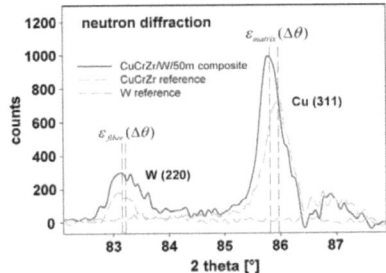

Fig. 4. A vertical sum over the PSD window shows Cu (3 1 1) matrix and W (2 2 0) fiber peaks acquired within 5 min.

$$\sigma_{matrix}^{trans} = \frac{E_{matrix}(T)\left(\varepsilon_{matrix}^{trans} + \upsilon_{matrix}\varepsilon_{matrix}^{long}\right)}{(1 + \upsilon_{matrix})(1 - 2\upsilon_{matrix})} \quad (2b)$$

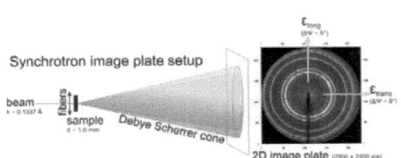

Fig. 5. Synchrotron diffraction, with monochromatic beam ($\lambda \sim 0.1337$ Å at 90 keV) in transmission: segmentation of the principal stresses by sum over $\Delta\psi = 5°$ of the 2D image delivering ε_{long} and ε_{trans} relative to the reference.

stresses were calculated with the same strain symmetry as used for the neutron diffraction according to Eq. (2c) and ($\varepsilon_{long} = (d - d_{trans})/d_{trans}$) to produce comparable results. Convergence of both neutron and synchrotron results obtained for W wires verified the uniaxial approach [16].

2.3. Synchrotron tomography

The highly brilliant synchrotron X-ray beam offers the possibility of high resolution synchrotron tomography. A white beam was used for absorption contrast imaging on ID15A at ESRF, Grenoble. The sample was rotated in the parallel beam setup acquiring ~1000 projections with a CCD camera behind. The high intensity of the polychromatic beam reduces the acquisition time down to less than 10 s. In situ 3D imaging of MFRMs during thermal cycling was made in-parallel to diffraction [17,18]. The CCD camera was moved between the 2D detector and the sample, alternating the acquisition between the tomography and diffraction.
Different CCD cameras were used: The Sarnoff CAM512 system with 512 x 512 pixels, 18 μm sensor pixel size for in situ scanning and the Dalsa Pantera1M60 system with 1024 x 1024 pixels, 12 μm sensor pixel size for the ex situ scans at RT (ID15A, ESRF). The matrix shell of the MFRM had been turned to suitable dimensions (d = 1.6 mm, l = 10 mm) for the in situ scans of simultaneous diffraction and tomography. The furnace setup offered a 180° free viewing angle for tomography. The synchrotron diffraction samples (d = 1.6 mm) and neutron diffraction samples (d = 3.5 mm) were used for the ex situ tomography scans on ID15A as well. Region of interest scans in the relatively large neutron samples produced some artefacts which could be corrected afterwards.

3. Results

Neutron diffraction measurements were made in the matrix of the Cu/SiC composites during thermal cycling using the Cu {3 1 1} peak. Scans at 0° and 90° to the fiber orientation delivered strain data in the two principal directions. Reference measurements on matrix samples of the same dimensions (d = 3.5 mm, l = 16 mm) as the composite samples were used to eliminate superimposed macro stresses possibly originated by gradients of fiber volume fractions and temperature. The acquired strain information provided matrix micro stresses averaged over the 70 mm³ gauge volume during thermal cycling. The results of a CuCrZr/SiC composite with 10, 15 and 30 vol.% of SiC (SCS6) fibers are shown in Fig. 6. Longitudinal and transversal matrix stresses during thermal cycling (RT – 550 °C) can be compared. Initially tensile stresses exist at RT in the matrix (Fig. 6b and c). Compression builds up in all samples during heating above ~300 °C by matrix expansion. During cooling, the stress inverts at ~400 °C again into tension which increases by further matrix shrinking as one approaches room temperature. Longitudinal tension is limited to ~350 MPa in the uncoated CuCrZr/SiC/10 m. The coated CuCrZr/SiC + Ti/ 15 m and CuCrZr/SiC + Ti/30 m reach 700 MPa longitudinal tension at RT. The second cycle behaves similarly to the first one. Microstress evolution during thermal cycling in Cu/SiC/20 m is shown in Fig. 7. Low initial tensile matrix stresses in Cu/SiC/20 m with pure Cu go into compression during heating and inverts back into tension at ~250 °C during cooling. Lower matrix stresses are built up in the uncoated Cu/SiC/20 m compared to the sample with the TiTaC interface. Longitudinal and transverse matrix stresses show similar behavior in all composites during thermal cycling, transverse being slightly lower in amplitude (Figs. 6c and 7c). The longitudinal stresses after 50 ex situ cycles are compared in Fig. 8 for pure Cu and CuCrZr matrix: The matrix stresses in CuCrZr/ SiC/10 m shift into tension; The stress amplitude stays constant

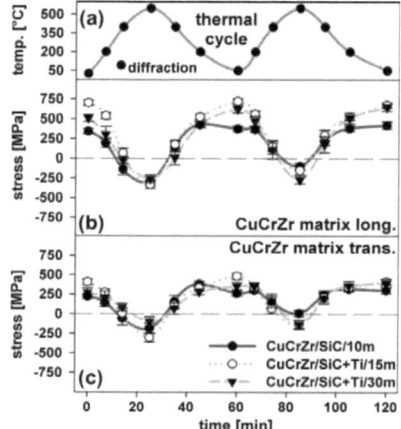

Fig. 6. Neutron diffraction: matrix micro stresses during two thermal cycles (RT – 550 °C) (a); longitudinal (b) and transversal (c) in CuCrZr (left) reinforced with SiC fibers of 10, 15 and 30 vol.%.

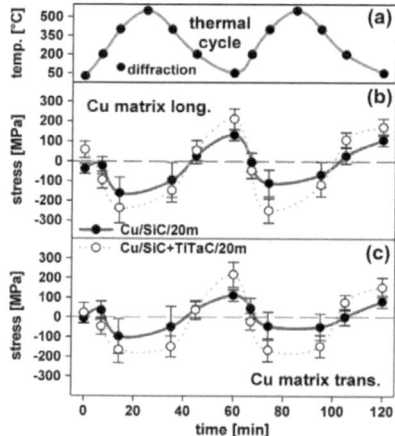

Fig. 7. Neutron diffraction: matrix micro stresses during two thermal cycles (RT – 550 °C) (a); longitudinal (b) and transversal (c) in pure Cu matrix reinforced with 20 vol.% SiC fibers.

in the Ti coated CuCrZr systems; In Cu/SiC/20 m the matrix stress amplitude stays almost constant after cycling; In Cu/SiC + TiTaC/ 20 m matrix stresses decrease significantly; Pure Cu matrix with TiTaC coated SiC fibers behave similar to CuCrZr matrix with uncoated SiC fibers.

Similar results are shown for W wire reinforced Cu/W/20 m composites in Fig. 9. The longitudinal stresses are higher in the graded interface system before further cycling, but suffer severe decay with a shift into tension after ex situ cycling. The etched wires produce low matrix stresses in both cases. W wire stresses could be determined in the Cu/W-graded/20 m due to good scattering properties of W with neutrons using the W {2 2 0} peak. The wire micro stresses in Cu/W/20 m reach up to ~400 MPa tension during heating. The wire stresses invert during cooling at

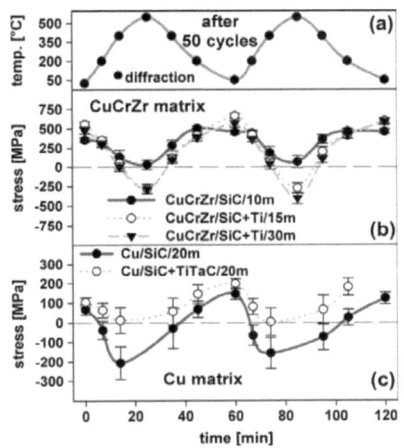

Fig. 8. Neutron diffraction: longitudinal matrix stresses during in situ thermal cycling (a) after 50 ex situ cycles (RT – 550 °C) in CuCrZr (b) and Cu matrix (c).

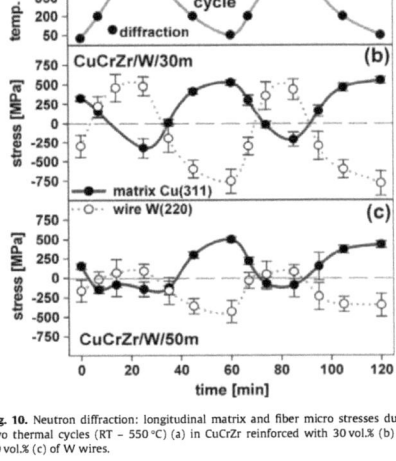

Fig. 10. Neutron diffraction: longitudinal matrix and fiber micro stresses during two thermal cycles (RT – 550 °C) (a) in CuCrZr reinforced with 30 vol.% (b) and 50 vol.% (c) of W wires.

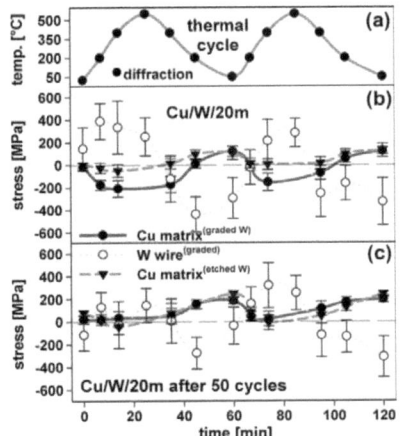

Fig. 9. Neutron diffraction of W wire reinforced pure Cu matrix: Longitudinal micro stresses during two cycles (a) before (b) and after (c) 50 ex situ cycles. Matrix stresses with graded and etched wire interface designs compared to fiber stresses.

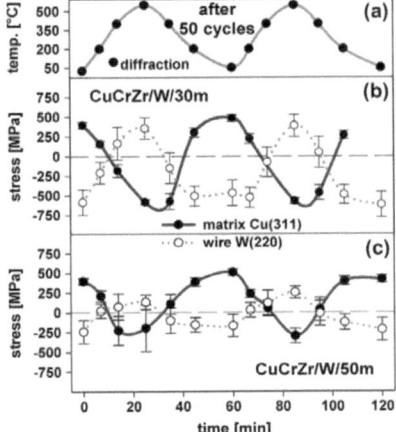

Fig. 11. Neutron diffraction: longitudinal matrix and wire micro stresses during two thermal cycles (RT – 550 °C) (a) in CuCrZr reinforced with 30 vol.% (b) and 50 vol.% (c) of W wires after 50 ex situ cycles.

~350 °C again into compression down to about −400 MPa at RT. After 50 ex situ cycles the stress amplitudes decrease both in the matrix and in the wires. Matrix and wire stresses in CuCrZr/W/ 30–50 m with 30 vol.% and 50 vol.% W wires are shown in Fig. 10. Higher stresses can be observed in the composite with 30 vol.% than in that with 50 vol.% wires. Matrix and wire stresses are complementary at the same level according to the wire's volume fraction. The initial residual stresses at RT increase when reaching RT again after cooling. Stresses in the same composites after 50 ex situ cycles (Fig. 11) are reproducible in the wires as well as in the matrix. Unfortunately, the SiC fiber micro-stress measurements did not succeed with neutron diffraction due to the weak scattering cross section of SiC but requiring short acquisition time to avoid matrix relaxation at high temperatures. Fig. 12 shows the results of synchrotron stress measurements in MFRM with coated and uncoated SiC fibers, Cu/SiC + Ti/20 m and Cu/SiC/20 m respectively. Matrix micro stresses in the uncoated system start in tension and change into compression during heating and invert into tension again after cooling. The SiC fiber stresses are of opposite sign compared to the matrix stresses compensating them in magnitude (micro-stress equilibrium). The fiber and matrix micro stresses are lower in the Ti coated MFRM sample due to initial fiber cracks identified in the gauge volume as shown in the corresponding tomographies in Fig. 12d and e.

The samples used for neutron diffraction are shown in Figs. 13 and 14 after 50 thermal cycles (RT − 550 °C). Different thermal fatigue damage types can be distinguished in the SEM images of transverse surfaces. SiC fiber and W wire reinforced pure Cu and CuCrZr matrix composites are shown in Table 2.

The anomalous elongation observed in some Cu/SiC/20 m samples after thermal cycling was further investigated. In Fig. 15a a metallographic cross section through a fiber of Cu/SiC + Ti/20 m shows one of the several fiber cracks. Matrix material is pressed into the crack between the fiber fragments. The same cracks and pores were identified by synchrotron tomography in Fig. 15b showing a single vertical slice of a 3D image. Fiber cracks open during heating to 550 °C but do not close completely after cooling again (Fig. 16). After two cycles the crack width increases 4 times its initial value. Strong interfacial damage after 50 thermal cycles appears in regions near fiber cracks shown in Fig. 17. Matrix voids and cracks are aggregating at the ends of the fiber fragments which get pushed apart. Voids in the matrix between the consolidated coated fibers are shown. These voids grow significantly during 50 thermal cycles (Fig. 18). Two types of matrix interface damage

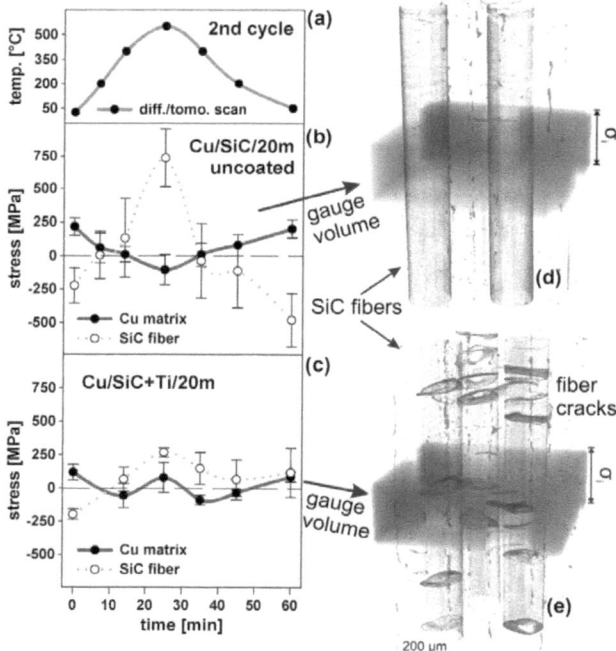

Fig. 12. Synchrotron X-ray diffraction longitudinal fiber and matrix micro stresses during thermal cycling (a) in a Cu/SiC/20 m with uncoated fibers (b), compared to a fragmented region in Cu/SiC + Ti/20 m (c). The tomography shows the gauge volume (d) and (e) corresponding to (b) and (c), respectively.

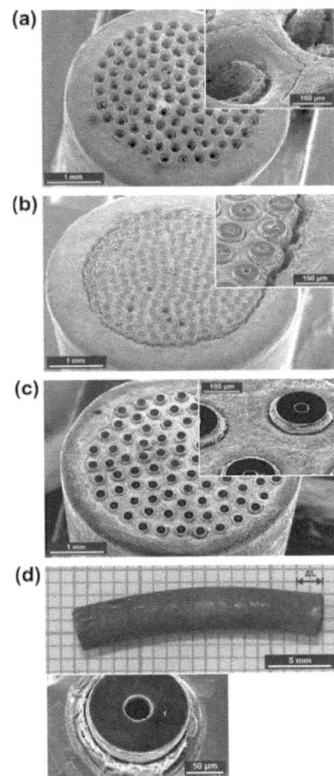

Fig. 13. SEM images (a–c) of SiC fiber reinforced Cu samples after 50 cycles (RT – 550 °C): (a) fiber intrusions in Cu/SiC/20 m, (b) matrix damage in Cu/SiC + Ti/20 m with fiber extrusions and (c) CuCrZr/SiC + Ti/15 m with fiber extrusions and no visible matrix damage. (d) light optical image of a Cu/SiC + Ti/20 m sample bent and elongated (ΔL – length change) after cycling up to 10% of initial length (d) with extruded fibers (SEM) on both ends.

can be distinguished: Radial cracks (Fig. 17b) and voids close to the interface with the fibers (Fig. 17c).

4. Discussion

In SiC monofilament reinforced Cu matrix composites internal stresses are generated during changing temperatures due to the CTE mismatch between the fiber and the matrix. These stresses lead to delamination of the interfaces and fiber sliding or matrix/fiber damage. The stronger the bonding the higher the micro stresses. Longitudinal micro stresses in a monofilament reinforced system during thermal cycling are given by Eq. (3) using the slab model [3].

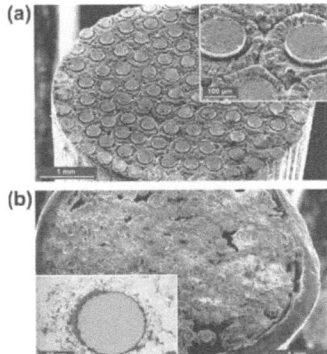

Fig. 14. SEM images of transverse surfaces of W wire reinforced copper samples after 50 cycles (RT – 550 °C): Stable shape of CuCrZr/W/30 m (a) with small surface extrusions. (b) severe matrix deformation and damage in the Cu/W/20 m with graded interface. Matrix cracks near interfaces in the LOM image of a cross section.

Table 2
Thermal fatigue damage types distinguished in the different composites.

Composite (m)	Damage type	Figure
Cu/SiC/20	Matrix deformation; matrix extruded/expanded over the fiber ends	Fig. 13a
Cu/SiC + Ti/20	Matrix damage between reinforced region and cladding; fibers extruded; some severely deformed samples; elongated up to 10% of its initial length	Figs. 13b and d and 15
CuCrZr/SiC + Ti/15	Little fiber extrusions; low matrix deformation	Fig. 13c
CuCrZr/W/30	No matrix damage; fibers well embedded	Fig. 14a
Cu/W-graded/20	Severe matrix deformation; matrix cracks near interfaces; W fibers remain bonded; no elongation; no extrusion	Fig. 14b

$$\sigma_{\text{matrix}} = E_{\text{matrix}}(\text{CTE}_{\text{MMC}} - \text{CTE}_{\text{matrix}})\Delta T \tag{3a}$$

$$\sigma_{\text{fiber}} = E_{\text{fiber}}(\text{CTE}_{\text{MMC}} - \text{CTE}_{\text{fiber}})\Delta T \tag{3b}$$

where σ is the stress, E is the Young's modulus, CTE is the linear coefficient of thermal expansion and ΔT is the temperature change.

The simplified linear elastic approach with ideal bonding is neglecting any transverse stress contributions. Strain controlled uniform longitudinal stress across each constituent is assumed. The CTE$_{\text{MMC}}$ calculated from the thermo-elastic model (Turner) includes the fiber volume fraction [11]. The different length change during changing temperatures of the matrix relative to its reinforcement has to be compensated by micro strains. The resulting CTE mismatch stresses (ΔT = 550 K) expected in the different composites are shown in Table 3.

Transverse to the fiber direction, the averaged stress level is a complex superposition of several contributions. Radial and Hoop (tangential) by transversal CTE mismatch, lateral contraction/expansion by Poisson mismatch and macro stresses by fiber distribution variations (Eq. (4)):

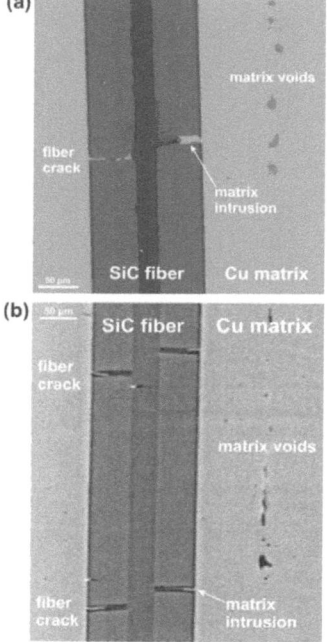

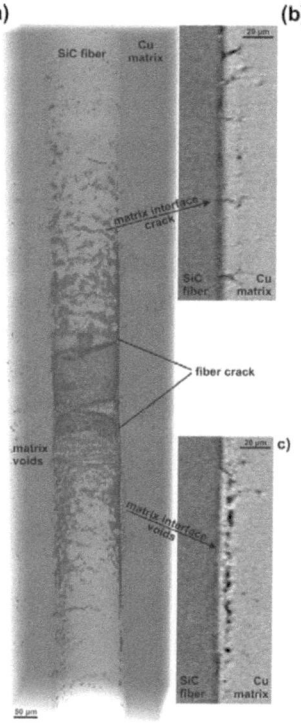

Fig. 15. Longitudinal sections through a fragmented SiC fiber with a crack in Cu/SiC + Ti/20 m and voids in the matrix: (a) scanning electron microscopic image after 50 cycles (RT – 550 °C) compared to (b) synchrotron tomography showing one image slice of a 3D scan (2 μm)³/voxel (ID15A) in the initial condition.

$$\sigma_{trans} = \sigma_{rad}(T) + \sigma_{Hoop}(T) + \sigma_{Poisson}(\sigma_{long}(T)) + \sigma_{macro}(\Delta v_f) \quad (4)$$

where Δv_f is the local variation in fiber/wire volume fraction.

The stress superposition from several components in transverse direction complicates the interpretation of stress correlations on bonding quality. To examine bonding strength of different composite types more emphasis was laid on the longitudinal contribution for comparison.

Fig. 17. Region in Cu/SiC + Ti/20 m around an initially fractured SiC fiber (a) after 50 ex situ cycles (RT – 550 °C), (1.6 μm)³/voxel (ID15A). Matrix voids in the bulk and at the interfaces (shearing) and matrix cracks can be distinguished. (a) 3D view, (b and c) longitudinal slices from a 3D reconstruction along fiber interface.

High matrix micro stresses are generated in CuCrZr along the SiC fibers during thermal cycling (ΔCTE ~ 13 ppm/K) with stress amplitudes up to ~800 MPa (Fig. 6b). The amplitude exceeds the calculated value enhanced by same signed transversal matrix

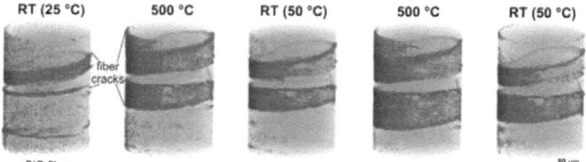

Fig. 16. Tomographic in situ observation of fiber cracks in Cu/SiC + Ti/20 m during thermal cycling (RT – 550 °C). The crack opens during heating but does not reversibly close during cooling.

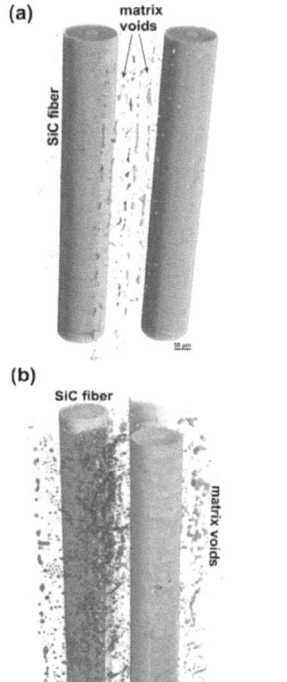

Fig. 18. Synchrotron tomography of SiC fibers in Cu/SiC/20 m with voids in the Cu matrix (transparent). The initial consolidation voids before thermal cycling (a) and after 50 ex situ cycles (b). The void volume fraction and sphericity increases.

stresses which maintain the longitudinal micro stresses. Hydrostatic stress components are responsible for stress levels higher than the matrix yield strength. The resulting uniaxial amplitude (assuming $\sigma_{trans} = 0$) would be significantly lower ($\Delta\sigma \sim 300$ MPa) similar to the prediction of the slab model (above). The initial value of ~ 500 MPa along the fibers at RT decreases during heating and inverts into compression of ~ -200 MPa at 550 °C. Softening and plastification of the CuCrZr matrix [6,7] at high temperatures occurs. The matrix stress is shifted into tension during cooling again. Ti coated fibers with high bonding strength reach the same high stresses independent of the fiber volume fraction (Fig. 6b). Matrix tensile stresses of uncoated SiC fibers in CuCrZr are limited to ~ 350 MPa during cooling due to fiber sliding. In pure Cu matrix composites the matrix stress amplitudes ($\Delta\sigma < 400$ MPa) are reduced by plastification of the soft matrix (Fig. 7). The SiC fibers without interface coating produce stress amplitudes in Cu of ~ 300 MPa predicted in Table 3. The strongly bonded TiTaC coated SCS0 fibers [8] generate high stresses above matrix yield strength due to multiaxiality.

After thermal cycling (50×, RT – 550 °C) the in situ matrix micro stresses decrease ($\Delta\sigma < 200$ MPa) for weakly bonded fibers in the strong matrix alloy and for strongly bonded fibers in a weak matrix alloy (Fig. 8). Debonded fibers in CuCrZr/SiC/10 m produce matrix tension during cooling limited to 350 MPa by fiber sliding. The matrix in Cu/SiC + TiTaC/20 m suffers severe deformation and damage due to the hydrostatic stress situation created by high fiber bonding strength. Matrix voids grow (Fig. 18) and interfacial radial matrix cracks and matrix voids are formed (Fig. 17).

W wire reinforced Cu behaves similarly to Cu/SiC/20 m. Matrix stresses decrease severely in strongly bonded Cu/W-graded/20 m ($\Delta\sigma < 350$ MPa to $\Delta\sigma < 150$ MPa) and little in weakly bonded Cu/W-etched/20 m ($\Delta\sigma < 200$ MPa) (Fig. 9) [9]. Decreasing microstress amplitudes are induced by severe matrix damage during thermal cycling. Inverse W wire stresses ($\Delta\sigma < 800$ MPa) compensate the matrix stresses. Both stress amplitudes decrease by thermal fatigue damage. The CuCrZr/W/30–50 m systems appear initially well bonded in the first two cycles and stay stable after 50 ex situ cycles (strong bonding with strong matrix in Fig. 10 and 11). Plastification in CuCrZr/W/50 m during the first cycle is caused by higher matrix stress contribution due to a lower matrix volume fraction. Residual stresses after cooling to RT vary with respect to the initial condition as a result of matrix plastification at elevated temperatures. These residual stresses at RT stay constant in the second and the following thermal cycles. Matrix stress amplitudes in CuCrZr/W/30 m (Fig. 10b) are similar to those in CuCrZr/SiC + Ti/30 m composites (Fig. 6b).

Fragmentation of brittle SiC fibers was observed in some strongly deformed samples after thermal cycling (Fig. 13d). Elongation of the samples up to 10% of their initial length is caused by severe matrix deformation with fibers pushed out on both ends. These fiber cracks originated from sample production and become partially filled with Cu matrix material during consolidation (Fig. 15b). The initial cracks open during thermal cycling by micro-stress relieve in the crack-near regions (Fig. 12). Crack opening (Fig. 16) and fiber pushing produces accumulating crack growth and macroscopic elongation of the composite into fiber direction with increasing number of cycles. After 50 cycles (RT – 550 °C) the matrix near these cracks gets severely deformed by high interfacial shear stresses suffering severe thermal fatigue damage (Fig. 17a): Matrix cracks (Fig. 17b) at the interfaces and voids (Fig. 17c) in the matrix are formed by shear stresses and creep during thermal cycling. Shear stresses [3] are the driving force for delamination near cracks, matrix deformation, fiber pushing and the resulting macroscopic elongation (Fig. 13d).

Table 3
Calculated fiber stresses in the constituents of different composites.

$\Delta T = 550$ K (m)	ν_{matrix}	E_{matrix} (MPa)	E_{fiber} (MPa)	CTE_{matrix} (ppm/K)	CTE_{fiber} (ppm/K)	CTE_{MMC} (ppm/K)	σ_{matrix} (MPa)	σ_{fiber} (MPa)
Cu/SiC/20	0.8	120	330	17	3	11	−376	1506
CuCrZr/SiC/30	0.7	128	330	16	3	9	−480	1121
Cu/W/20	0.8	120	410	17	5	11	−365	1459
CuCrZr/W/30	0.7	128	410	16	5	10	−448	1045
CuCrZr/W/50	0.5	128	410	16	5	8	−590	590

5. Conclusions

- Micro stresses higher than matrix yield strength are generated in monofilament reinforced copper composites during thermal cycling (CTE mismatch) due to hydrostatic components. The better the bonding strength the higher the stresses are.
- Fiber sliding was observed in MFRM with uncoated SiC fibers. Ti coatings improve the bonding strength of SiC fibers significantly.
- Even uncoated W wires achieve good bonding strength with Cu matrix. High matrix and wire stresses are built up in Cu/W composites during thermal cycling. High matrix stresses cause more damage in pure Cu than in CuCrZr.
- High bonding strength generates severe damage in a weak matrix material. Pure Cu suffers ductile deformation at high temperatures and damage during cooling by tension. Matrix voids (shear) and radial matrix cracks (partial delamination) are formed at the interfaces of strongly bonded wires during thermal cycling.
- Precipitation hardened CuCrZr reduces damage at high temperatures by softening, at low temperatures by higher fracture toughness. The fiber/wire stress is relieved during heating by relaxation of the matrix.
- SiC fiber cracks formed during sample manufacturing produce severe accumulation of thermal fatigue damage in crack-near regions during cycling. Stress induced shear, fiber pushing and matrix deformation cause a macroscopic elongation of the composite in fiber direction.
- Good bonding strength reduces delamination, but increases matrix stresses. High micro stresses need a strong matrix to withstand alternating thermal loads under service conditions.
- Ti coated SiC fibers or W wires well bonded in a tough CuCrZr matrix are promising materials for high temperature heat sink applications (sliding < bonding strength < matrix damage). Cracks in brittle SiC fibers have to be avoided during production.
- Neutron diffraction with its high penetration depth allowed non-destructive measurements of type 2 micro stresses with good statistics in relatively big gauge volumes (especially in W containing materials).
- Synchrotron tomography helps to understand thermal fatigue damage propagation by 3D in situ damage visualization with high resolution.

The tested Cu/W and Cu/SiC composites are currently irradiated at NRG in Petten, Netherlands at temperatures and dpa comparable to their operation condition. Afterwards the MFRM samples will undergo further tensile tests and creep tests to investigate the long term effects of irradiation on the stability of the SiC fiber and W wire reinforced copper composites.

Acknowledgements

The research was financed by the ExtreMat 6th framework EU project, and the authors would like to thank all the cooperating partners especially J. Hemptenmacher, DLR Köln, Germany and A. Herrmann, IPP Garching, Germany. Last but not least we would like to thank the helpful support from the staff of HZB Wannsee, Germany and ESRF Grenoble, France helping, organizing and carrying out successful experiments on their sites. M.S. especially acknowledges fruitful discussions with the chairwoman of UGLI.

References

[1] E. Becvar, Aspekte der Kernfusionsforschung, Verlag der Österreichischen Akademie der Wissenschaften, Informationstagung April, 1986.
[2] Panayiotis, J. Karditsas, Fusion Eng. Des. 30 (1995) 307–323.
[3] T.W. Clyne, P.J. Withers, An Introduction to Metal Matrix Composites, Cambridge University Press, Trumpington street, Cambridge CB2 1RP, 1993.
[4] J.H. You, H. Bolt, J. Nucl. Mater. (2002) 305.
[5] A. Brendel, C. Popescu, T. Köck, H. Bolt, J. Nucl. Mater. (2007) 367–370.
[6] P. Peters, J. Hemptenmacher, D. Muchilo, Phys. Scripta T128 (2007) 209–212.
[7] P. Peters, J. Hemptenmacher, H. Schurmann, Comput. Sci. Technol. 70 (9) (2010) 1321–1329.
[8] T. Köck, A. Brendel, H. Bolt, J. Nucl. Mater. (2007) 362.
[9] A. Herrmann, K. Schmid, M. Balden, H. Bolt, J. Nucl. Mater. (2009) 386–388.
[10] A. Brendel, V. Paffenholz, T. Köck, H. Bolt, J. Nucl. Mater. (2009) 386–388.
[11] P. Turner, J. Re. Nat. Bu. Stan. 36 (1946) 239–250.
[12] P. Peters, J. Hemptenmacher, H. Schurmann, Mater. Sci. Forum (2007) 931–935.
[13] M.E. Fitzpatrick, A. Lodini, Analysis of Residual Stress by Diffraction using Neutron and Synchrotron Radiation, Taylor & Francis, London, 2003.
[14] R.C. Wimpory, P. Mikula, J. Saroun, T. Poeste, J. Li, M. Hofmann, R. Schneider, Neutron News 19 (1) (2008) 16–19.
[15] V. Hauk, Structureal and Residual Stress Analysis by Nondestructive Methods, Elsevier Science B.V. Sare Burgerhartstraat 25, Amsterdam, NL, 1997.
[16] M. Schöbel, J. Jonke, H.P. Degischer, A. Herrmann, R.C. Wimpory, T. Buslaps, Adv. Eng. Mater. (2011).
[17] R. Sinclair, M. Preuss, E. Maire, J.Y. Buffiere, P. Bowen, P.J. Withers, Acta Mater. 52 (2004).
[18] M. Schöbel, W. Altendorfer, H.P. Degischer, S. Vaucher et al., Internal stresses and thermal fatigue in SiC particle reinforced aluminum composites for heat sink applications, Comput. Sci. Technol., submitted for publication.

Acknowledgments

To all the people I would like to thank for their important contribution to this work that were not mentioned in the included publications:

First of all I thank Peter Degischer who committed me project responsibility of very interesting and challenging work on his institute. The commitment of this topic chosen for my PhD thesis is gratefully acknowledged.

In same degree I owe a debt of gratitude to Rainer Schneider and Tobias Poeste for their great support of my work. Their assistance during my very first proposed neutron experiments as inexperienced scientist was no less than enabling this thesis.

Great thanks have to be addressed to Guillermo Requena, Domonkos Tolnai and Heinz Kaminski who introduced me into several practices and sequences which make synchrotron experiments successful and even possible. Their experienced support during several experimental campaigns was an important criterion for the success of this work.

I am also deeply indepted to Heidemarie Knoblich for several laboratory sample preparation and metallographic characterization sessions. The mechanical and chemical manipulation of the composite materials that has been very demanding required her expertized knowledge.

Christian Linsmeier and Frantisek Simancik supported my work in the ExtreMat project. Their professional organization allowed good cooperation between several materials manufacturers and our institution which was essential for the variety of characterized materials.

Last but not least I thank Elisabeth Magerl and Elisabeth Zerbst for their help during the final phase of this thesis. Very fruitful substantial as well as conceptional critics and discussions helped me to accomplish this work.

Die VDM Verlagsservicegesellschaft sucht für wissenschaftliche Verlage abgeschlossene und herausragende

Dissertationen, Habilitationen, Diplomarbeiten, Master Theses, Magisterarbeiten usw.

für die kostenlose Publikation als Fachbuch.

Sie verfügen über eine Arbeit, die hohen inhaltlichen und formalen Ansprüchen genügt, und haben Interesse an einer honorarvergüteten Publikation?

Dann senden Sie bitte erste Informationen über sich und Ihre Arbeit per Email an *info@vdm-vsg.de*.

Sie erhalten kurzfristig unser Feedback!

VDM Verlagsservicegesellschaft mbH
Dudweiler Landstr. 99   Telefon +49 681 3720 174
D - 66123 Saarbrücken   Fax   +49 681 3720 1749
www.vdm-vsg.de

Die VDM Verlagsservicegesellschaft mbH vertritt

Printed by Books on Demand GmbH, Norderstedt / Germany